AF552469

CHEMISTRY OF ENZYMES

ENCYCLOPAEDIA OF ENZYMOLOGY

Vol. II

CHEMISTRY OF ENZYMES

By

Dr. Arvind N. Shukla

School of Studies of Zoology & Biotechnology

Vikram University

Ujjain

DISCOVERY PUBLISHING HOUSE PVT. LTD.

NEW DELHI-110 002

First Published-2008

ISBN 978-81-8356-357-4 (Set)

Published by:

DISCOVERY PUBLISHING HOUSE PVT. LTD.

4831/24, Ansari Road, Prahlad Street,
Darya Ganj, New Delhi-110002 (India)
Phone: 23279245 • Fax: 91-11-23253475
E-mail: dphbooks@rediffmail.com
dphtemp@indiatimes.com
Website: www.discoverypublishinghouse.com

Printed at:

Sachin Printers, Delhi

Preface

This classic *Encyclopaedia of Enzymology* has been compiled to provide an understanding of enzymology for university degree study and teaching. It comprehensively covers theory and applications of these biological life supporting catalysts in a readable book that will serve as a bridge to more advanced and specialized areas. It not only introduces the fundamentals and biochemistry of enzymes but also provides the molecular and experimental background. This unique piece of work compiled five volumes containing wealth of information for researchers, microbiologists, and biotechnologists. This excellent title is intended mainly for students taking degree course which have a substantial biochemistry component. The chief aim of this particular book is to help student understand the concepts involved in enzymology. An attempt has been made to give a perspective of each topic, and examples are quoted where appropriate.

There can be no claim to originality except in the manner of treatment and much of the information has been obtained from the books and scientific journals available in the different libraries.

The author express his thanks to his friends and colleagues whose continue inspirations have initiated him to bring out this book.

The author is painfully aware of the shortcomings, errors and misprints that have crept in, and shall be grateful to receive suggestion for improvement of the next edition from all the readers.

The author express his gratitude to Mr. Wasan and staff of M/s Discovery Publishing House Pvt. Ltd. for their whole hearted co-operation in the publication of this book.

Author

Contents

Chapter 1 1—10

INTRODUCTION

Secretion, Transport of Enzymatic Proteins, Pancreatic Secretion, Hypothetical Views, Duality, The Assumptions, Testing a Constructed Hypothesis, Specific Hypothesis, Logical Empiricism, Constructed Hypotheses, A Constructed Paradigm

Chapter 2 11—29

ENZYMES

Introduction/History, Classification and Nomenclature, Chemical Energetics, Thermodynamics, Kinetics, How Enzymes Work, Transition State Stabilization, Catalysis by Approximation, Covalent Catalysis, Acid-Base and Redox Catalysis, Enzyme Kinetics, Michaelis-Menten Equation, Regulation, Nontraiditional Enzymes, Perspectives, Enzymes, High Temperature, Introduction, Sources of High Temperature Enzymes, Properties of High Temperature Enzymes, Structural Information on High Temperature Enzymes

Chapter 3 30—38

ENZYME ASSAYS

Units of Enzyme Activity, Initial Rates, Why Does the Rate Decrease?, Substrate Depletion, Approach to Equilibrium, Product Inhibition, Inactivation of Enzyme, Artifactual Cause, Types of Assays, Continuous or Discontinous Assay?, Direct Assays, Indirect Assays, Coupled Assays, Estimation of Initial Rate, Linearity of Measured Rate with Enzyme Concentration

Chapter 4 39—53

TRANSPORT

Electron Microscopy, Attendant Problems, The Importance of Specific Mechanistc Hypotheses, Structure-Function Hypotheses,

Cellular Permeability, Vesicle Formation, Assumptions, The First Assumption, Semipermeability, Membranous Pores, Cellular Aggregation

Chapter 5 **54—64**

MOTOR PROTEINS

Cellular and Subcellular Dynamics, Families of Motor Proteins, Motors Proteins, Background and Discovery, Diversity Among the Motors Proteins, Cellular Roles Played by Motor Proteins, Future Perspectives

Chapter 6 **65—78**

EXOCYTOSIS

Existential Hypothesis, Falsification of Hypotheses, Evidence of Form, Secretory Granules in Cells, Disappearance of Granules, Exocytosis, Membrane Surface Area, Particles, General Discussion, Exocytosis in the Pacncreas

Chapter 7 **79—83**

ENZYME MECHANICS

Theoretical Aspects, Methods, Examples

Chapter 8 **84—112**

SECRETORY GRANULES

Filling of the Granules, Vesicles from the RER, Complexity of the RER, Mathematic Solution, Smooth-surfaced Vesicles, Condensing Vacuole, Dynamics of Fusion, Condensing Vacuole, Filling of Condensing Vacuoles, Membrane Flow, Disappearing Zymogen Granules, Constancy of Number, Granule Turnover, Disappearance of Granules and Secretion of Protein, Granule Numbers and Volume Density, Granule Volume and the Amount of Enzyme Secreted, Blockage Secretion

Chapter 9 **113—117**

COENZYMES

General History, Individual Coenzymes, Vitamin-Derived Coenzymes, Other Coenzymes

Chapter 10 **118—125**

ISOENZYMES

Derivation of Isoenzymes, Structure, Function, and Developmental

Aspects, Methods of Analysis, Practical Applications of Isoenzyme Analysis, Nonisoenzymic Molecular Forms of Enzymes

Chapter 11 126—132

CYTOCHROME

Characteristics of P450s, Activities of P450s, How Many P450s Are There?, Regulation of P450s, Future Directions

Chapter 12 133—138

PHOSPHOLIPASES

Phospholipids, Phospholipase Specificity, Phospholipase A2, Secretory Phospholipase A2, Cytosolic Phospholipase A2, Phospholipase C, Phospholipase D

Chapter 13 139—144

RIBOZYME CHEMISTRY

Genetic Information, Catalytic Function of RNA, Self-Splicing Rna, Self-Cleaving Catalytic RNA, Ribonucleoprotein Enzyme

Chapter 14 145—184

CYTOPLASMIC ENZYMES

Cytoplasmic Proteins, Specific Activity, Intracellular Enzymes, Physiology of Digestive Enzymes, Anatomical Approach, Functional Cytoplasm, Osmotic Effect

Chapter 15 185—235

SYNTHESIS VERSUS TRANSPORT

Trypsin Inhibitor, Synthetic and Transport Effects, Parallel Secretion, Parallel transport, Laboratry Opinion, Nonparallel Transport, End-Product of Digestion, Transport of Chymodenin, Independent Secretion, Fluctuation, Energetics, Borders of the Paradigm

Chapter 16 236—267

ENZYMES OF PROTEIN METABOLISM

Enzyme Data, Specificity, Molecular Mass, Isoelectric Point, pH Optimum, Assay, Stability, Inhibitors, Experimental Procedures, Nucleic Acid Isolation, Protein Studies, Carboxypeptidase Y, The Enzyme, Purification, Specificity, Physical and Chemical Properties, pH Optimum, Assay, Stability, Denaturing Agents, Inhibitors, Additional Comments, Experimental Procedure,

Determination of C-Terminal Sequences in Peptides and Proteins, Aminopeptidases, Enzyme Data, Pyroglutamate Aminopeptidase (Calf Liver), Stability, Aminopeptidase M, Prolidase (Porcine Kidney), Experimental Procedures, Removal of Pyroglutamic Acid, N-Terminal Sequence Determination by Time-Course Hydrolysis with Aminopeptidase M, The Use of Aminopeptidase M, Alkaline Phosphatase, The Enzyme, Bacterial Alkaline Phosphatase (BAP), Calf Intestinal Alkaline Phosphatase (CLAP), Enzymic Reaction, Substrate, Temperature, pH, Cations, Inhibitors, Sulfhydryl Reagents, Enzyme Assay and Unit Definition, Experimental Procedures, Storage and Stability, Reaction Conditions, Reaction Protocol, Polynucleotide Kinase, The Enzyme, Enzymic Reaction, Reaction Catalyzed, Substrate, Acceptor, Temperature, pH, Cations, Activators, Inhibitors, Sulfhydryl Reagents, Enzyme Assay and Unit Definition, Experimental Procedures, Uses of Polynucleotide Kinase, Storage and Stability, Reaction Conditions, Reaction Protocols

INDEX 268—274

1

INTRODUCTION

There are Two biological processes-the secretion of organic products by cells and the transport of one of these products, proteins, across biological membranes. Understanding of each of these processes has become entwined with that of the other. This intermingling has historical, intellectual, and experimental roots whose basis and character will form an important and recurring theme throughout this book.

SECRETION

Secretion can be broadly divided into two types. The first is secretion in which substances, mostly fluid and electrolytes, move across polarised tissue surfaces, particularly epithelia, from one extracellular compartment to another, namely, from blood or the internal milieu to bodily cavities or the external milieu. Its opposite is *absorption*. The second type of secretion is the release or efflux of organic compounds from cells that manufacture or accumulate them.

Only this second type of secretion will concern us here and my use of the term "*secretion*" will refer exclusively to this, process. Such secretion processes are widely seen, diverse in terms of the biological products secreted, and imposing because of their important place in biological function. They account for most, if not all, communication between cells, as in the secretion of *neurotransmitters* and *hormones*; in the secretion of digestive enzymes, they account as well for the ability of organisms to digest food, a function central to the survival of all animal species.

When we say that a substance is secreted by a cell, we imply the existence of an underlying biological *purpose* or *function* for the process, such as neurotransmission or digestion. Sometimes a purpose is assumed even though it is unknown, that is, it awaits discovery, or why else would the substance be secreted? Of course, although the release of a contained substance from a cell may serve some biological function, its release in and of itself does not demonstrate it, and the activity may be purposeless. A more convincing inference about function can be made when an externally applied stimulus, particularly a "*physiological*" one such as a neurotransmitter or hormone, is found to increase the rate of release of a specific substance. In this case a function is implied by the *response,* although even here increased release may merely reflect a passive, dependent, and purposeless sequel to another process.

Thus, *secretion* is often thought of as a special process in which the transport of a substance is thought to serve a particular function or purpose, and while this inference may sometimes be justified (that is, be more than teleology) and sometimes is not, nevertheless, secretion is often set apart in this way from biological transport as a general phenomenon in which the mere transport of a molecule tells us nothing of purpose.

The term secretion is also used as a synonym for the model that attempts to explain the transfer of "*secreted*" molecules across cell membranes, namely, the vesicle paradigm. Although this model is complicated and varies in its details from system to system, the central and final step in the construct, the step that is thought to lead directly to the release of the product from the cell, is common to them all. It proposes the fusion of the membrane of intracellular vesicles, which contain or are thought to contain the products that are secreted, and the cell membrane; the result is a hole produced in both membranes through which vesicle contents leave the cell and enter the extracellular environment. This model, commonly called *exocytosis*, will have an important place in the discussion and analysis that follows.

Thus, the study of secretion is frequently set apart from the general study of biological transport in two important ways. First, there may be a purpose, known or assumed, for the process—a particular raison d'etre. And second, release may be thought to occur in a particular manner. In this way, ideas of *purpose* and *mechanism* have often formed an a priori assumptive foundation upon which knowledge of particular secretion processes is built. Unlike the general study of biological transport, our understanding of secretion is not usually derived from a description of the events as free from special, as opposed to general, mechanistic, and functional hypotheses as possible, but is often developed in just such a context. Indeed, preconceived notions of mechanism and function have often taken the place of an actual or real description of the system under study, undertaken *without* presumed particular purposes or presumed mechanisms of transport. The question of whether the proverbial cart-models of mechanism and hypotheses of purpose-has been placed before the horse-a full and unbiased description of the transport events as they occur—shall appear frequently in what follows. I will try to separate the concept of the *process secretion* from specific hypothetical models that attempt to explain how it is carried out, as well as from particular functions in which it may be involved.

TRANSPORT OF ENZYMATIC PROTEINS

Protein transport, the second process that will concern us here, includes examples of secretion, such as *peptide hormone* and *digestive enzyme* secretion, as well as thc more general topic of the permeability of cells and biological membranes to large molecules which includes transport processes that we would not characterize as "*secretory*"; for example, protein transport across intracellular membranes (mitochondria, chloroplasts, peroxisomes, and so forth), the uptake of peptidic bacterial toxins by host cells, and the absorption of proteins across intestinal epithelium.

Protein transport and the secretion of organic cell products are processes that exist at different levels of logical organisation. In the case of secretion, we consider different classes of molecules being transported out of cells by a process called secretion with whatever functional or mechanistic trappings we might wish to attach to that word, whereas protein transport is the movement of a particular class of molecules across biological membranes in general whatever orientation, location, mechanism, or purpose such transport might have.

Our understanding of each of these processes has had an important bearing on how we have come to view the other. This is primarily because of a particular experimental focus over the course of a century on *how cells secrete proteins*. Because the secretion of protein is an example of both sets (formally, the class "*protein secretion*" forms a subset at the intersection of the two larger sets "*secretion*" and "*protein transport*"), our view of the secretion of organic molecules has been greatly influenced by our understanding of how cells secrete the specific class of molecules, proteins, and similarly, our knowledge of protein secretion has been the substrate on which our understanding of protein transport has developed.

PANCREATIC SECRETION

The two general mechanisms are *secretion* and *protein transport*, by means of a particular member of the subset protein secretion. One reason for this choice is that, as protein secretion has been the process upon which much of our knowledge of both secretion and protein transport has been built, our knowledge of protein secretion has in great part been derived from this specific case, *the secretion of digestive enzymes by ducted glands, in particular the mammalian pancreas.*

HYPOTHETICAL VIEWS

During this century, *reductionism* has come into full flower not only as the *philosophical* underpinning of biology, but in a sense as its central experimental methodology. Based on the idea that biological forms appear to be made of the same stuff as the rest of the material world and obey the same laws, *reductionism* as an experimental strategy argues that we can understand their workings through knowledge of their parts, ultimately at the fundamental level of the substituent molecules and atoms. From this atomistic knowledge the nature of the native or whole system can then be inferred. This experimental reductionism and the philosophical system that proposes that biological law can be "*reduced*" to more general natural laws are not one in the same as is so often thought. Nonetheless, the experimental approach has been strikingly successful. We have made great progress in defining the contents of cells and organisms both *anatomically* and *chemically*. What would probably be agreed upon as the major achievement of 20th century biology, the determination of the structure of DNA and the details of the *genetic code*, is no doubt the brightest star in the firmament of experimental reductionism in biology. Nevertheless, and in no way demeaning the importance of these achievements, merely knowing, however intimately, the parts of a system provides us with only half of the information that we must have in order to understand how the intact system works. If our models are to be satisfactory, then we must also have a substantial independent description of the workings of that intact system, a four-dimensional space-time continuum, into which the presumed parts must fit. To describe a system we must have knowledge of its parts, obtained from atomistic scientific investigation *and* an accurate and equally rigorous description of the system's behavior in its native state, that is, how the parts interact with each other as part of the observed natural continuum.

Trying to construct a process solely from knowledge of its parts, or presumed parts, is somewhat like trying to solve a four-dimensional puzzle in which the pieces can be fit together in numerous ways, without knowing whether you have all of the pieces, or do not have extra pieces mixed in from another puzzle, and most importantly without knowing the nature of the picture that you are trying to construct. To contradict this view, one might point to the elucidation of the structure of DNA. After all, didn't the genetic code fall out of our knowledge of this structure? But what could one have inferred about the function of the DNA molecule, knowing its structure completely, without the work of geneticists over the course of the prior 75 years defining the nature and behavior of the gene? That is, considerable knowledge of the whole genetic system fortunately *preceded* the discovery of the structure of DNA.

The great changes that have occurred during this century in our view of the physical world came about in great part because mechanical *Newtonian models* could not account for certain observations, particularly those related to the nature of the physical continuum. If we believe that we can explain the workings of biological systems in physical terms, then we must understand both the parts and the whole, that is, the molecules, reactions, and organelles that comprise the whole, *and* the way in which the native system functions. To do otherwise is to ignore one important aspect of the physical world, while focusing solely on another. Physicists have tried to explain both the parts (matter) and the continuum (the context in which matter exists) within a single theoretical framework. The issue is the same in biology.

DUALITY

Some physiologists argue that one learns nothing (or precious little) from in vitro, or worse yet, cell-free, studies because we can only learn from the whole system with all of its natural functions in place, and that all else is artifact. On the other side, there are those who just as vigorously argue that the complexities of the whole system in vivo are beyond rational analysis, and that "*rigorous*" science can only be carried out on parts of systems and cells.

In order to understand, we must know of what the system is comprised and how the isolated parts behave, as well as about the intact process. Awareness of this necessary duality seems to be lacking when mechanistic models are developed solely or primarily from knowledge of the parts of a system and their isolated behavior. It will be called this a "*constructionist*" approach to model building in which the researcher constructs a model for an intact system based on his or her knowledge of its parts in the absence of a substantial *independently* arrived at description of the intact system. Of course, it is only knowledge of its presumed parts, since they can only be presumed in this case. There is nothing wrong with constructing mechanistic hypotheses based on knowledge of a system's parts when there is also knowledge of the native system with which to confront a constructed model, as in the case of DNA, or, even in the absence of such knowledge as a starting place for further investigation and an ordering of thoughts—a working hypothesis, if you will. However, hypotheses constructed in the absence of an adequate understanding of the native system can pose difficulties that go beyond the substantial uncertainties of trying to put together an extremely complex and unknown puzzle.

THE ASSUMPTIONS

When a model is not derived from the known characteristics of a natural system, it can only be constructed on the basis of *assumptions* about that system; that is, how we think it should behave. These assumptions may either be a matter of convenience or belief: convenience—assumptions simply made to facilitate the experimental process—or belief—in that certain views are held as being generally applicable, either in the absence of evidence, or on the basis of evidence from other systems. In experimental biology such assumptions are rarely akin to assuming that some universal theorem holds, such as energy and matter being conserved. Nor are they similar to mathematical assumptions that help define a theoretical system without necessarily judging its claim to being true. In the absence of significant knowledge of the native process, that is, its *kinetic* and *thermodynamic* characteristics, assumptions used in constructing hypotheses and models for biological processes are likely to be based on evidence collected in other systems that is deemed applicable to the system of interest; or even sometimes on beliefs that are not supported by substantial experimental evidence or other proof at all. In the latter case, we often only have our intuition or feelings as a guide. While in the former, we must necessarily also hold (assume) that the observation in the other system is correct, that the interpretation given in that setting is correct, and that the analogy is apt. For these reasons, such assumptions are often weak and clouded by chains of uncertainty, which are then reflected in the constructed hypothesis.

TESTING A CONSTRUCTED HYPOTHESIS

Constructed hypotheses cannot be tested against knowledge of an intact process; that is, we cannot ask whether or not our model is consistent with the known behavior of a natural process, because its behavior is unknown to us. But could we not argue that while this criticism of the constructed hypothesis may have some merit, it should not pose any serious problems for the hypothesis used as a tool in the design of experiments? Indeed, isn't that really the purpose of our hypothesis—to permit us to learn about the workings of the natural process in an orderly way? Therefore, shall we not inevitably learn about these things as we proceed with experimentation? Although we may learn about the intact process by testing our hypothesis, there is no guarantee that this will be an outcome. Testing a system in the context of a particular model may only provide information about the system's natural behavior if the proposed model turns out to be correct; that is, the experiments may add to our knowledge of the system's real behavior if and only if our hypothesis is correct, and not regardless of whether or not it is correct. Thus, it is possible to test a particular hypothesis in many different ways and yet be essentially as ignorant of the native process.

SPECIFIC HYPOTHESIS

But even if this were not the case, and our tests clearly provided information about the real system, we might still have substantial problems if our hypothesis were a relatively specific and complex mechanistic construct at the outset.

When complex hypotheses are proposed for processes in the absence of substantial knowledge of the system's characteristics in the native state, tests of specific aspects of the hypothesis (subordinate hypotheses of the larger model) can at best only provide us with an extremely limited view of the natural continuum. That is, what we learn about it is necessarily derived from a selected and limited sampling of its behaviors as seen through the myopic eye of our hypothesis. The broader or more general our hypothesis, the less restricted our probing and the more we can hope to learn from it. The more specific and particular our predictions, the less we will learn about the system overall. This may seem somewhat perverse in a "reductionist" context since the latter type of experiment often appears to be more rigorous, getting to the essence of "mechanism" because of its concrete and detailed character. One may even talk of molecules or parts of molecules and their role in the larger process. But nevertheless it seems to me that the more general our hypothetical statement, the more valuable the experiment and the broader the resultant understanding of the *natural system*.

LOGICAL EMPIRICISM

If our hypothesis fails a test, then it will be rejected and replaced by a new one; and that with time and experimental test we shall nevertheless hone in on the true nature of the system? This approach to experimental science, often called *logical empiricism*, rests to a considerable degree on the belief that the hypothesis will be rejected as a result of a negative test, and will be replaced by another that will then similarly be tested against reality. *Hypotheses* are usually derived from a *scientific paradigm* deeply embedded in an area and based on a wide range of evidence, assumptions, and related beliefs. A negative test of such a construct, by means of a particular hypothetical prediction derived from it, often, and perhaps usually, leads, at least initially, to the rejection of the negative observation, not the paradigm.

Moreover, the likelihood that a construct will be rejected by a negative test becomes increasingly remote the more specific and complex the model. In this case, negative evidence is often readily explained by means of ad hoc or *auxiliary hypotheses* that represent minor modifications of the comprehensive construct, and is often attributed to the investigator's own, as yet, incomplete understanding of the details of the construction. This allows one to hold on, sometimes with great tenacity, to the model's general concept despite negative evidence. On the other hand, the more general a hypothesis, the more likely its rejection will be required by experimental test, and hence its greater value.

CONSTRUCTED HYPOTHESES

What happens if tests of the hypothesis are positive? It is of course in the nature of a hypothesis that if experimental support for it is obtained, its claim on reality will be strengthened. Sometimes so are the assumptions that gave rise to it in the first instance, however circular this reasoning may be. Thus, as the researcher gains support, however tentative, for his or her particular hypothesis, the model as a tool in the design of experiments, gives way to a specific construct that attempts to describe real processes concretely, and the claim may even be made that the proposed process or mechanism actually occurs as envisioned. Making this leap may be problematic, however, particularly if we have used the most common logical methodological tool in experimental biology, weak inductive inference, to test our hypothesis.

In this case, we make a prediction about the behavior of a system based on our model, a hypothesis deduced from it, if you will, and then test this hypothesis. If the prediction is proven correct, then we infer that our general hypothesis is correct as well. However, this result really only permits us to say that "if our hypothesis is correct, then the following should happen, and it does." It does not rule out alternative explanations for the same phenomenon along which path the truth may lie instead. It is rare, if it ever occurs in the world of experimental biology, that we are capable of saying "if and only if our hypothesis is true will the following occur."

Nevertheless, when a series of tests of this sort are carried out and a chain of positive answers provided, the tentative quality of the original hypothesis can be lost in a most elating affirmation that the investigator's ideas were correct. Such lines of evidence can be very deceiving as tests in the real world since observations that are consistent with false hypotheses are rather common occurrences. Thus, a list of such observations, however long, can never alone prove that our hypothesis is true, because it does not provide a unique solution for the observations, nor consider one explanation against another. This problem is accentuated when our model has been constructed without an independent description of the natural continuum against which to test it.

This does not mean that we should discard experiments whose design leads to inferences of the weak inductive variety. Indeed, it is hard to imagine experimental biology in their absence. And even if models derived from such weak inferences cannot be considered as satisfactory approximations of reality, since science is an unending search for better and better approximations, they may be useful for the moment whatever their shortcomings. The problem occurs when we lose sight of the fragility of hypotheses supported by this type of evidence alone; once again, particularly when our hypothesis is for a system for which we do not have substantial knowledge of the intact process other than as viewed from within the framework of our specific hypothesis.

A CONSTRUCTED PARADIGM

In the absence of an independent description of the natural process, our hypothesis and our description of the process may be made of the same cloth. The parts of the system that we describe are only those that our hypothesis leads us to, and sometimes not even that. The more specific and elaborate our hypotheses become, the more likely we are to go wrong, because our models are easily modified in trivial ways to account for negative observations without the whole construct being seriously challenged. Thus, in the absence of independent knowledge of the behavior of the whole native or natural system, we cannot properly test such hypotheses in their own right, or in comparison to others.

2

ENZYMES

Enzymes are the biological catalysts (usually proteins) that help carry out nearly all chemical reactions in living systems; indeed, life as we know it would not be possible without them. Enzymes are true catalysts: that is, they increase the rate of a reaction but do not participate in it; they are returned to their original form at the end of the reaction cycle. Enzymes are special as catalysts only because their specificity and rate enhancements are unparalleled by any man-made or chemically developed catalysts. Enzymatic rate enhancements can range from 10^3 to 10^{16} times the rates of uncatalyzed reactions. This ability allows otherwise exceedingly slow chemical reactions to occur on a time scale that is biologically meaningful. Like other catalysts, enzymes lower the activation energy of a reaction, making the product kinetically accessible.

The science of enzymology has made remarkable strides in the last 50 years, largely because of advances in recombinant DNA techniques, X-ray crystallography, various types of spectroscopy, and the ready availability of isotopes for kinetic and structural studies. Our understanding of enzymatic mechanisms of catalysis is increasing every day. Continued study of enzymes may lead to the development of supercatalysts for use in industry, in bioremediation of toxic waste, and as therapeutic agents for the treatment of disease. Many diseases are the result of a missing or defective enzyme; research in this area is bringing us closer to the ability to repair or create substitutes for these essential molecules.

INTRODUCTION/HISTORY

Enzymes are responsible for carrying out the highly complex

chemical reactions necessary to sustain life. Like chemical catalysts of nonbiological origin, enzymes only increase the rate of a reaction that would naturally occur; they cannot induce a reaction that would not ordinarily happen. For example, in the absence of enzymes, the starch in last night's baked potato would break down into simpler sugars, but the process would take hundreds if not thousands of years. In the presence of some of your digestive enzymes, the job is done by the time of this morning's drive to work-an enzyme has increased the rate of a spontaneous reaction. No enzyme, however, could catalyze the conversion of simple sugars to starch (in a potato, for example) without the input of additional energy; this is because the process would not occur naturally. Another feature that enzymes have in common with nonbiological catalysts is the regeneration of the catalyst at the end of the reaction; enzymes are not consumed in the course of a reaction and are returned to their initial form by the end of the catalytic cycle.

Enzymes may be *highly* specific for the reactions they catalyze, selectively binding their substrates and efficiently converting them to different chemical forms, the products. Often an enzyme will selectively bind one substrate out of the many thousands of chemicals in a cell, and increase the rate of one specific reaction by millions of times. Other enzymes are much less specific in their *binding,* but the reaction catalyzed is still highly specific. For example, the mammalian cytochrome P450 binds a variety of substrates, but in each case the chemical reaction is the addition of a hydroxyl group (—OH) to the substrate.

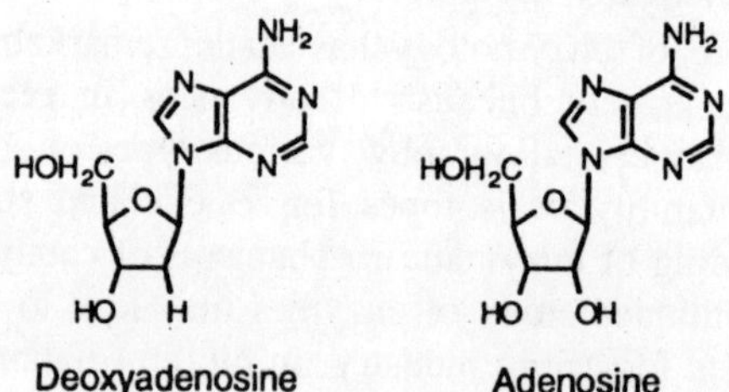

Figure 2.1 : Enzymes can easily distinguish between these two very similar molecules.

Specificity is a major feature that distinguishes enzymes from nonbiological catalysts. Another feature that distinguishes enzymes from chemical catalysts is regulation. Enzymes may be regulated in a variety of ways that may increase or decrease their catalytic efficiency. Finally, enzymes may be distinguished from most chemical catalysts by the phenomenon of saturation. Enzymes have a maximum rate with respect to substrate concentration, whereas the majority of chemical catalysts do not.

Table 2.1 : A Brief History of Enzymology

Date	Investigators	Contribution
1833	Payen and Persoz	First observation of enzyme activity in a test tube
1878	Kilhne	Introduces the word "enzyme" (in yeast)
1898	Duclaux	First suggestion for enzyme nomenclature
1890s	Fischer	"Lock and key" model of enzyme action
1913	Michaelis and Menten	Mathematical model of enzyme action
1920s	Willstatter	First pure enzymes
1926	Sumner	First crystallization of an enzyme (urease)
1948	Pauling	Proposed transition state theory of enzyme action
1951	Pauling and Corey	Discovery of a-helix and p-sheet structures in enzymes
1953	Sanger	First determination of the amino acid sequence of a protein (insulin)
1961	Perutz and Kendrew	X-Ray structure of myoglobin (not an enzyme)
1986	Cech	Discovery of catalytic RNA
1986	Lerner and Schultz	Catalytic antibodies developed

Some of the landmark events in the history of enzymology. Throughout much of this history, all enzymes were believed to be proteins-polymers of amino acids, the so-called building blocks of life. However, it has recently been shown that ribonucleic acids (nonprotein biopolymers) can catalyze certain reactions as well, and as biological catalysts, these RNA molecules fit the old definition of an enzyme. Most biochemists now have narrowed the definition of the term "enzyme" to refer only to proteins and have accepted the term "ribozyme" to describe catalytic RNA molecules. Another recent development has been the discovery of the means to produce antibodies that can catalyze specific reactions. These catalytic antibodies can also be regarded as enzymes, although not all the reactions they catalyze are of biological significance.

CLASSIFICATION AND NOMENCLATURE

Enzymes are given names that help to classify them. The agency responsible for naming enzymes is the Commission on Biochemical Nomenclature of the International Union of Biochemistry, which periodically publishes updates to a manuscript entitled *Enzyme Nomenclature*. There are three acceptable ways to refer to an enzyme: its Enzyme Commission code number, its systematic name, and its recommended name.

The EC code number provides a rigid systematic method for classifying enzymes. It takes on the general form of four numbers, separated by periods, following the letters "EC." For example, EC 1.1.1.1 represents alcohol dehydrogenase. The number in the first position of the code number denotes the class of the enzyme. This information indicates the general type of reaction the enzyme carries out. The second number is the enzyme subclass. The third indicates the sub-subclass. It is important to realize that the subclass and sub-subclass designations mean different things in different classes. The last number is the enzyme's serial number. Enzymes within a given sub-subclass are assigned sequential serial numbers, giving each enzyme a unique code.

CHEMICAL ENERGETICS

Thermodynamics

To understand how enzymes work, we must first examine the energetics of chemical reactions. There are two major sets of properties common to all chemical reactions: kinetic and thermodynamic. In a simple sense, the kinetic parameters of a reaction describe how fast the reaction (or chemical steps within the reaction) will take place. Thermodynamics, on the other hand, indicates the extent to which reactants will be converted to products. Thermodynamic predictions rely on the relative stabilities of products and reactants.

Enzymes affect *only* the *kinetics* of the reaction, since they, like all other chemical catalysts, cannot alter the equilibrium ratio of products to reactants. The most useful thermodynamic value for determination of whether a reaction will occur is the change in Gibbs free energy (ΔG), named for the American chemist J. W. Gibbs.

The general relationship between the change in Gibbs free energy and other thermodynamic parameters for any chemical reaction is as follows:

$$\Delta G = \Delta H - T\Delta S \tag{1}$$

where ΔG is the change in free energy, ΔH is the change in enthalpy, T is the temperature in degrees kelvin, and ΔS is the change in entropy. Enthalpy is a measure of stored energy and entropy, in a broad sense, is the degree of disorder of a system.

All chemical reactions are theoretically reversible, such that reactants are constantly being re-formed from products. The rates of the forward and backward reactions depend on the concentrations of the reactants and products as well as on "rate constants" for the forward and back-reactions, which are different for every reaction. The higher the concentration of reactants, the faster the forward reaction will go; the higher the concentration of products, the faster the reverse reaction will go. Since the reactants are depleted as they become transformed into products during the course of a chemical reaction, the rate of the forward reaction is decreasing and the rate of the reverse reaction is increasing.

Eventually, the reaction will reach a state of equilibrium; at equilibrium there is no *net* formation of products or reactants-products are converted to reactants and reactants are converted to products at equal rates.

To determine whether a reaction will occur, it is useful to look at the equilibrium constant (K_{eq}) for the reaction. For the chemical reaction shown in equation (2),

$$A + B \rightleftharpoons C + D \tag{2}$$

the equilibrium constant is defined as follows

$$K_{eq} = \frac{[C][D]}{[A][B]} \tag{3}$$

where [X] indicates molar concentration of X at *equilibrium*. If the ratio of the initial concentrations of products to reactants do not satisfy the ratio defined by K_{eq}, the reaction will proceed spontaneously in whichever direction is necessary to achieve the equilibrium ratio K_{eq}. It is important to realize that a change in any of the initial concentrations will alter the concentrations of reactants and products needed at equilibrium to insure that their ratio is still equal to K_{eq}.

Now that we have defined the equilibrium constant, its relationship to the Gibbs free energy can be shown:

$$\Delta G = \Delta G^{\circ} + RT \text{ In} \left(\frac{[C][D]}{[A][B]}\right) \tag{4}$$

where ΔG° is the standard free energy change, a constant for a given chemical reaction; R is the ideal gas constant, and T is the temperature in degrees kelvin.

By definition $\Delta G = 0$ at equilibrium, so the relationship becomes:

$$\Delta G^{\circ} = -RT \text{ In} K_{eq} \tag{5}$$

The importance of ΔG, specifically ΔG°, is evident. Since ΔG° is directly related to the equilibrium constant, it can be shown that any reaction that will proceed in the forward direction will have a negative ΔG. Armed with the initial concentrations of reactants and products and with the value of the constant ΔG° (or with the value of K_{eq}, which can be used to calculate ΔG°), it becomes an easy task to predict the direction of the reaction (i.e., whether more C and D will be formed or more A and B). The value of ΔG° is proportional to the negative log of the equilibrium constant (equation 5), and therefore indicates the extent to which the reaction will favor products over reactants at equilibrium.

Two points should be noted:

1. The ΔG for a reaction changes with the concentrations of reactants and products. A positive ΔG for a reaction mixture does not mean that no reaction will occur; the positive sign simply means that the *forward* reaction will not go spontaneously under that set of conditions. In fact, a positive ΔG value means that under those conditions, there will be a net back-reaction. A change in the initial concentrations or temperature may make the forward reaction favorable.

2. The value of ΔG indicates nothing about how fast a reaction will proceed. The ΔG can provide information only about the direction and extent of the reaction that will bring the reaction to equilibrium under a certain set of conditions. Even though the products of a reaction may be greatly favored at equilibrium (large K_{eq}), it could take many, many years for that state to be reached.

KINETICS

We now turn our attention to the rates of chemical reactions. The rate of most chemical reactions is dependent on the concentrations of the substrates (reactants). For the reaction shown in equation (6),

$$A + B \rightleftharpoons C + D \tag{6}$$

the forward rate (v) can be expressed as the product of the concentrations of A and B times a rate constant *k:*

$$v = k\,[A][B] \tag{7}$$

Thermodynamics does not provide information about the rate of a reaction. By using transition state theory, however, we can obtain the rate law for this reaction. The most convenient way to apply the properties of chemical reactions just described to the explanation of rates and catalysis is to construct a reaction coordinate diagram. Reaction coordinate diagrams for an uncatalyzed and an enzyme-catalyzed reaction. Since the final products of the reaction possess less free energy (lower free energy is a more stable, thermodynamically favored state), thermodynamics tells us that the product is favored at equilibrium. However, the reactants must first assume a very unstable, high energy state called the transition state. The free energy needed to promote a molecule (or molecules) from the ground state to the transition state is called the activation energy (denoted OG$). As the reaction proceeds from left to right in the uncatalyzed instance, a large activation energy must be overcome.

Transition state theory assumes that the reactants (A and B) in a chemical process are in equilibrium with the transition state $\left(AB^{\ddagger}\right)$. Because of this equilibrium, we may apply the formula for Gibbs free energy to the activation energy, giving:

$$\Delta G^{\ddagger} = -RT\ \text{In}\left(\frac{[AB]^{\ddagger}}{[A][B]}\right) \tag{8}$$

Statistical thermodynamics allows us to calculate the probability that a molecule will have a particular energy. If we take ΔG as that energy, we obtain the probability of a molecule having enough energy to reach the transition state:

$$p = \frac{\kappa T}{h}\exp\left(-\frac{\Delta G^{\ddagger}}{RT}\right) \tag{9}$$

where the new parameters are κ, the Boltzmann constant, and h, Planck's constant.

It is obvious that the rate of a reaction is governed by the number of molecules that have enough energy to reach the transition state of the *slowest* step in the reaction, the rate-determining step. Assuming that once the transition state has been reached, half the molecules will go forward to products and half will go back to reactants, we can express the rate law of the reaction as follows:

$$v = \frac{\kappa T}{2h}\exp\left(-\frac{\Delta G^{\ddagger}}{RT}\right)[A][B] \quad (10)$$

and therefore the rate constant is given by t

$$k = \frac{\kappa T}{2h}\exp\left(-\frac{\Delta G^{\ddagger}}{RT}\right) \quad (11)$$

which depends only on $\Delta G^{\ddagger}$ and the temperature.

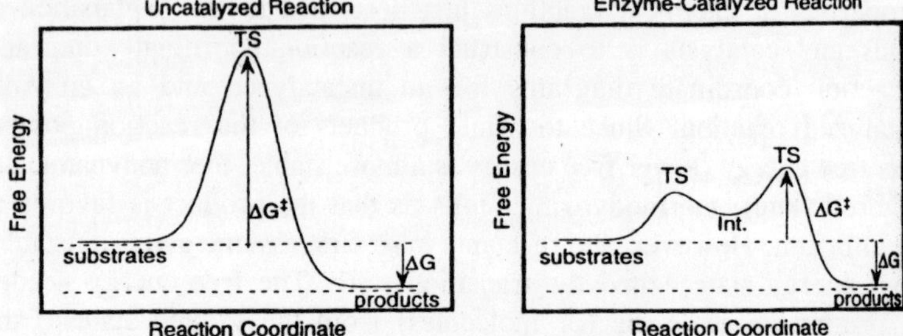

Figure 2.2 : Reaction coordinate diagrams for uncatalyzed and catalysed reactions. This uncatalysed reaction has a single transition state, but some reactions may have more than one. The catalysed reaction has two transition states, indicating that catalysed reactions need not have the same mechanism as uncatalysed ones. TS: transition state. ΔG: free energy of the reaction. ΔG: activation energy of the reaction. Int.: an intermediate in the reaction.

Since very few molecules possess the energy needed to attain the transition state, the magnitude of ΔG provides the major kinetic obstacle in any reaction. The reaction coordinate diagram of an enzyme-catalyzed reaction shows the alternate chemical pathway with a more accessible, lower free energy transition state (and thus a larger rate constant). It should be noted that the relative stability of products and substrates remains unchanged, and the rate of the reverse reaction is also enhanced by the same factor as the forward reaction.

HOW ENZYMES WORK

Enzymes enhance reaction rates by factors of 10^3 to 10^{16} relative to the rates of the uncatalyzed reactions. Some enzymes (e.g., carbonic anhydrase, triosephosphate isomerase, catalase) are considered to be "catalytically perfect" because the rate of the reaction is limited only by the rate at which the substrates diffuse through water to encounter the catalyst. Enzymes achieve these phenomenal rate enhancements by lowering the reaction's activation energy. To accomplish this, enzymes use a variety of strategies, including stabilization of the transition state, catalysis by approximation, acid-base and redox catalysis, and covalent catalysis.

Transition State Stabilization

The most important aspect of enzymatic catalysis is the ability of the enzyme to stabilize the transition state of a reaction and thereby lower the activation energy of the overall reaction. Enzymes are able to selectively bind the transition state more strongly than the ground state of a molecule because they contain a highly evolved binding site that is complementary in structure to the transition state of the reaction rather than the ground state of the reactants. Since the enzyme is designed to bind the transition state, when the substrate binds the substrate is "bent" into a conformation that is closer to the transition state and thereby lowers the energy necessary to attain the transition state. The enzyme may be complementary to the transition state in size, shape, and charge distribution. The transition state fits into the enzyme as a key fits into a lock, and binds tightly. For efficient catalysis, however, the enzyme needs to be able to "turn over" (i.e., to complete one reaction cycle) as quickly as possible; thus the enzyme should not bind the *product* very tightly.

Catalysis by Approximation

For molecules to react in the uncatalyzed reaction, they must collide with one another in a very precise orientation. In contrast, when a molecule is bound to an enzyme, it is held in an optimal orientation and proximity for reaction with another chemical group. The randomness, or entropy, involved in obtaining the correct reactive conformation is greatly reduced when the reactants are bound to the enzyme, resulting in a smaller $\Delta G^{\ddagger}$.

Covalent Catalysis

The chemical side chains of an enzyme's constituent amino acids present in the active site may actually take part in the catalyzed reaction, and be restored to their original state by the end of the reaction cycle. A commonly observed mechanism is one in which an amino acid side chain displaces a chemical group of a substrate molecule to form a covalently bound enzyme-substrate intermediate, which is more reactive than the original substrate. A second substrate may then react with the intermediate complex to yield product(s) and the enzyme in its original, catalytically active form. This strategy is well precedented in hydrolysis and other transfer reactions. In these cases, the enzyme provides an alternative reaction mechanism with a smaller $\Delta G^{\ddagger}$.

Acid-Base and Redox Catalysis

Amino acid side chains may also cause a chemical reaction to occur more readily by acting as acids (donors of protons) or bases (acceptors of protons). It is important to note that other (nonprotein) acids or bases can also act as catalysts, but in enzymes, these reactive groups are oriented in the active site so that the necessary proton transfer may occur most efficiently.

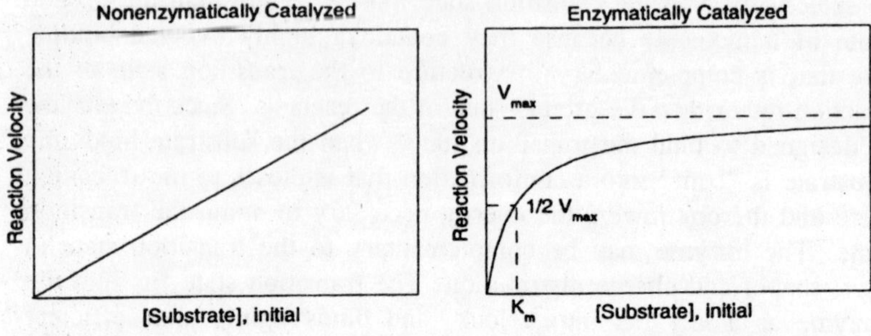

Figure 2.3 : Effect of substrate concentration on the initial velocity of reactions catalysed nonenzymatically and enzymatically.

Electrons may also be transferred during the course of a reaction; such reactions are called *redox* (reduction-oxidation) reactions. Reduction occurs when a molecule gains electrons and oxidation occurs when electrons are lost. Enzymes most commonly accomplish these electron transfers by incorporating transition metals or specialized organic molecules called *coenzymes*. Metals and coenzymes are generally referred to as *cofactors* and may also take part in acid-base chemistry and/or in covalent catalysis. Again, there are nonenzymatic redox catalysts, but the active site of an enzyme is situated so that everything is in the correct orientation to optimize catalytic efficiency.

ENZYME KINETICS

One of the characteristics distinguishing enzymes from most nonenzymatic catalysts is the phenomenon of *saturation*. A plot of the initial velocity of an enzyme-catalyzed reaction versus the initial concentration of the substrate indicates that the rate of the reaction varies linearly with low substrate concentrations; the reaction *is first-order* (varying linearly) with respect to substrate concentration. However the rate soon asymptotically approaches a limiting value, called V_{max}, becoming *zeroorder* with respect to substrate concentration (not varying with concentration). By comparison, a similar plot typical for a reaction catalyzed nonenzymatically shows no such limit; it is first-order with respect to substrate concentration over the entire range of substrate concentrations that can be obtained.

Michaelis-Menten Equation

Henri, Michaelis, and Menten developed a mathematical treatment to explain substrate saturation, which was extended by Briggs and Haldane. Consider a simple enzyme-catalyzed reaction as depicted in equation (12) in which substrate (S) binds to enzyme (E) in a reversible equilibrium and is then converted into a product (P) with the regeneration of the enzyme.

$$E + S \underset{k_1}{\overset{k_1}{\rightleftharpoons}} ES \xrightarrow{k_2} E + P \qquad (12)$$

Here we define the concentration of the enzyme-substrate complex to be [ES], the concentration of the free enzyme to be [E], and the total enzyme concentration to be $[E_0]$ (= $[E]$ + $[ES]$). The concentrations of substrate and product are represented by [S] and [P], respectively. The rate constants for the two forward reactions are indicated by k_1 and k_2, while the reverse reaction is indicated by k_1.

Mathematically, a first-order rate is expressed as a rate constant times a concentration. For example, the rate expression for the conversion of *ES* to *E* and *P* would be written as $k_2[ES]$, the rate constant times the concentration of *ES*. The rate of *change* of the concentration of the intermediate ($d[ES]/dt$) is equal to the rate of its formation from *E* and *S*, minus its rate of breakdown. There are two modes of breakdown, one to *E* and *S*, the other to *E* and *P*. The overall rate of change of the concentration of *ES* is therefore given by:

$$\frac{d[ES]}{dt} = k_1[E][S] - k_1[ES] - k_2[ES] \qquad (13)$$

Briggs and Haldane introduced the concept of the ***steady state***. The steady state assumption postulates that although initially the rate of production of *ES* would be positive, because no *ES* is initially in solution, very shortly after *E* and S have been mixed, the rate of formation of *ES* equals the rate of its breakdown. Therefore $d[ES]dt = 0$, implying a steady state, so that

$$\begin{gathered} k_1[E][S] - k_1[ES] - k_2[ES] = 0 \\ k_1\{[E_0] - [ES]\}[S] - k_{-1}[ES] - k_2[ES] = 0 \end{gathered} \qquad (14)$$

Solving for $[ES]$, one obtains:

$$[ES] = \frac{k_1[E_0][S]}{k_{-1} + k_2 + k_1[S]} \qquad (15)$$

Since the rate of the overall reaction is the rate of formation of P from *ES*, the rate expression is $v = k_2[ES]$. The rate expression then becomes:

$$v = \frac{k_1 k_2 [E_0][S]}{k_{-1} + k_2 + k_1 [S]} \tag{16}$$

Finally, substituting k_0 for k_2, where k_0 may be a combination of several rate constants, not simply a single step as indicated in our example, and setting $K_m = (k_{-1} + k_2)/k_1$, one arrives at the Michaelis-Menten equation:

$$v = \frac{k_0 [E_0][S]}{k_m + [S]} \tag{17}$$

This equation, which describes a rectangular hyperbola, explains the substrate saturation behavior characteristic of enzyme-catalyzed reactions. One notices that as the substrate concentration $[S]$ becomes very large ($\gg K_m$) the value of $K_m + [S]$ may be approximated by $[S]$. The rate equation then reduces to:

$$v = k_0[E_0] \tag{18}$$

explaining the zero-order behavior at high concentrations of substrate. The maximal velocity, V_{max} is thus mathematically equal to $k_o[E_o]$.

At $[S]$ equal to K_m, the rate equation becomes:

$$v = \frac{V_{max}}{2} \tag{19}$$

Therefore the K_m value is not only a ratio of rate constants as defined earlier; it is also the substrate concentration necessary to obtain half of V_{max}.

Regulation

Another distinguishing feature of enzymatic catalysis is that it can be regulated. An enzyme may respond to changing conditions within a cell by slowing down or speeding up its activity as necessary. This is usually accomplished by the binding of a small molecule to the enzyme. This molecule changes the conformation of the enzyme in such a way as to make it either more or less efficient, depending on the needs of the cell. Enzymes that are regulated in this manner still follow Michaelis-Menten kinetics, albeit slightly modified.

A second form of regulation of enzyme activity is the phenomenon of *allostery,* meaning "other site." Allosteric enzymes are frequently composed of multiple subunits. The essence of allostery is that ligand binding to one subunit induces conformational changes in that subunit and also in the other *subunits,* altering the enzymatic properties of *the other subunits*. A "*ligand*" in this case may be a substrate or an allosteric effector (either an activator or inhibitor). The kinetic behavior of allosteric enzymes is frequently so complex that it cannot be described by the Michaelis-Menten equation. Some allosteric enzymes are regulated by binding of the allosteric ligand(s), altering the apparent affinity of the enzyme for substrate, a property that is reflected in the K_m. Allosteric regulation of other enzymes occurs at the level of catalytic efficiency and results in a change in V_{max}. Binding of allosteric ligands to some enzymes will alter both V_{max} and K_m.

NONTRAIDITIONAL ENZYMES

For many years it was thought that all enzymes were the highly evolved proteins described in the preceding discussion. In 1986 reports were published of two new types of biological catalyst: catalytic antibodies and catalytic RNA.

The catalysis of reactions by antibodies is not a natural occurrence. Antibodies are protein molecules that bind foreign substances as part the immune response. Antibodies are designed to bind the ground states of molecules. Scientists were able to produce catalytic antibodies by taking advantage of transition state theory, using a molecule similar in structure to the proposed transition state of the desired reaction to elicit an antibody that can catalyze the reaction. The production of catalytic antibodies by this method is a major validation of the hypothesis that transition state stabilization contributes significantly to reaction rate enhancement by enzymes.

Catalytic RNA molecules, or *ribozymes,* discovered by Thomas Cech, are the first known example of nonprotein biological catalysts. Ribonucleic acid is another kind of biological polymer that is heavily involved in the synthesis of proteins using the genetic information stored in a cell's DNA. RNA and DNA have different characteristics; the most important difference for catalysis is the ability of RNA to assume complex three-dimensional structures. Biological catalysts must possess the ability to bind a specific substrate; the three-dimensional folds in an enzyme or an RNA molecule allow selective substrate binding. Catalytic RNA molecules are able to cut and splice themselves into a form that can then act catalytically on other RNA molecules. To date, the only ribozyme-catalyzed reactions known are cleavages of RNA and DNA. Ribozymes are true catalysts because they enhance reaction rates without any net change to themselves. They are capable of turnover (recycling) and show kinetics typical of enzymes.

A very interesting aspect of the discovery of catalytic RNA lies in the evolutionary implication. It is conceivable that RNA molecules were in existence before DNA or proteins. It has been proposed by

Walter Gilbert that RNA molecules first catalyzed their own replication and then developed other catalytic abilities. It is not difficult to envision that RNA molecules eventually developed the ability to synthesize proteins. Proteins, which were able to catalyze many more reactions with greater efficiency than ribozymes, became the primary biological catalysts. It is speculated that DNA was first synthesized by reverse transcription of the genetic information on RNA and that DNA became the carrier of genetic information because it is more stable than RNA.

PERSPECTIVES

Enzymology has a long and exciting history. Largely because of development of recombinant DNA technology, however, we are now able to isolate in pure form large amounts of numerous enzymes that could not be isolated from the native biological source, and to alter the catalytic activity of natural enzymes. Recent improvements in physical methods, including NMR spectroscopy and X-ray diffraction from crystals of large molecules, are allowing dramatic advances to be made in our understanding of the mechanisms and modes of communication between enzymes and other biological macromolecules. It would appear that the glory days of enzymology are not behind us, but before us.

ENZYMES, HIGH TEMPERATURE

In the last few years, microorganisms have been discovered that grow optimally at temperatures near and above 100°C, the normal boiling point of water. Most of these so-called hyperthermophiles have been found in marine volcanic environments, which include deep-sea hydrothermal vents. They are the most ancient organisms known. The majority are strict anaerobes and depend on the reduction of elemental sulfur (S) to hydrogen sulfide (H_2S) for optimal growth. Most of them utilize proteins, peptides, and sugars as growth substrates, which are converted to simple organic acids and gases such as hydrogen (H_2) and carbon dioxide (CO_2). These potentially rich sources of a variety of "hyperthermostable" enzymes are able to catalyze reactions at extreme temperatures. So far about 30 enzymes have been purified and characterised from the sulfur-reducing species. All have optimal temperature for catalysis above 100°C and some have half-lives at this temperature of several days.

INTRODUCTION

A revelation of great significance in the field of microbiology, with profound implications in microbial metabolism, biochemistry, and biotechnology, occurred in 1982 when microorganisms that grew at temperatures exceeding 100°C were isolated from shallow marine volcanic vents. Although we know little about the novel biochemistry that must be required to sustain these organisms under such conditions, they are potential sources of a range of "*hyperthermostable*" or high temperature enzymes. This contribution discusses what is known about hyperthermophilic organisms and the enzymes that have been characterised from them, with an emphasis on the enzymes from the "*sulfur-dependent*" species, the predominant type of hyperthermophile.

SOURCES OF HIGH TEMPERATURE ENZYMES

Hyperthermophiles are defined here as organisms able to grow at 90°C and above with an optimal growth temperature of at least 80°C. They have been isolated in just the last few years, and only two of the 20 or so hyperthermophilic genera currently known are conventional bacteria. The majority of them are classified as Archaea (formerly Archaebacteria). Archaea were recognised as the third kingdom of life in the late 1970s on the basis of molecular (rRNA) sequence analyses. These studies also indicate that hyperthermophilic organisms are the most ancient of life forms, the first to have diverged from some universal ancestor, which suggests that the rest of biology is the result of evolutionary pressures to adapt to low (< 100°C) temperatures.

The known hyperthermophilic genera are listed in Table 1. All have been isolated from geothermally heated environments, which include deep-sea hydrothermal vents located up to 4000 m below sea level. The majority of the hyperthermophiles are referred to as sulfur-dependent organisms, since their growth depends to a greater or lesser extent on the reduction of elemental sulfur (S°) to hydrogen sulfide (H_2S). Almost all grow only under strictly anaerobic conditions and are strict organotrophs that use complex organic mixtures as the carbon and nitrogen sources, including yeast and meat extracts, tryptone, peptone, and casein. A few are also able to use carbohydrates as additional or primary C sources. They are potential sources of a variety of hydrolytic-type enzymes, such as proteases and amylases, which are of some industrial importance.

PROPERTIES OF HIGH TEMPERATURE ENZYMES

The enzymes that have been purified from the S°-dependent

hyperthermophilic genera are listed in Table elsewhere in this chapter. All exhibit maximal catalytic activity above 100°C. Protease activity has been detected in several hyperthermophiles, but these organisms have been purified from only two species. The serine-type protease of *Pyrococcus furiosus* was isolated by boiling cell-free extracts with the detergent sodium dodecyl sulfate (SDS, 1%) for 24 hours, a process that destroyed virtually all other cellular proteins. Several carbohydrate-

Table 2.2 : Hyperthermophilic Genera

Genus	T_{max} (°C)	Growth Substrates	Habitat
S°-Dependent Archaea			
Thermoproteus	92	Pep, CBH, H_2, S^0	c
Staphylothermus	98	Pep, S^0	d/m
Desulfurococcus	90	Pep, S^0	d/c/m
Thermofilum	100	Pep, S^0	c
Pyrobaculum	102	Pep, H_2, S^0	c
Acidianus	96	S°, H_2/S^0, O_2	c/m
Pyrodictium	110	Pep, CBH, H_2, S^0	d/m
Thermodiscus	98	Pep, S^0	m
Pyrococcus	105	Pep, CBH, ± S^0	d/m
Thermococcus	97	Pep, CBH, ± S^0	d/m
Hyperthermus	110	Pep, H_2, ± S^0	m
"ES-I"	91	Pep, CBH, S^0	d
"ES-4"	108	Pep, CBH, S^0	d
"GB-D"	103	Pep, S^0	d
"GE-5"	102	Pep, S^0	d
"JDF-3"	108	Pep, CBH, S^0	d
Sulfate-reducing Archaea			
Archaeoglobus	95	CBH, H_2, SO_4	d/m
Methanogenic Archaea			
Methanococcus	91	H_2, CO_2	d/m
Methanothermus	97	H_2, CO_2	c
Methanopyrus	110	H_2, CO_2	d/m
Bacteria			
Thermotoga	90	Pep, CBH, ± S^0	d/m
Aquifex	95	S^0, H_2, O_2, NO_3	m

metabolizing enzymes able to utilize starch, maltose, cellulose, and/or xylan as substrates have been characterised. These represent the most stable enzymes yet characterised from any organism, with half-lives of one to two days near 100°C. The activities of *Pyrococcus woesei* amylase, Ps *furiosus* sucrose n-glucohydrolase, and ES-4 amylopullulanase have been measured at 130, 117, and 135°C, respectively. Similarly, the immobilised xylanase from a *Thermotoga* strain retained 25% of its activity after I hour at 130°C in molten sorbitol.

Many of the hyperthermophiles also metabolize hydrogen gas (H_2), and hydrogenase is the enzyme responsible for catalyzing its production and activation. Hydrogenases usually utilize as an electron carrier a low molecular weight redox protein known as ferredoxin, which also has been purified from these organisms. Another low molecular weight redox protein known as rubredoxin has been purified from *P. furiosus* as well, although its function in the cell is not known. The rubredoxin is of some importance, since it is the only hyperthermophilic protein for which a threedimensional structure is available (determined independently by NMR spectroscopy, X-ray crystallography, and molecular modeling). Surprisingly, the overall folding patterns of the mesophilic and hyperthermophilic rubredoxins are remarkably similar, and enhanced stability appears to arise from rather minor changes.

Several ferredoxin-dependent oxidoreductases have also been purified. These oxidize various substrates and transfer electron to ferredoxin ultimately for H_2 production or S° reduction. The aldehyde oxidoreductase of *Ps furiosus* converts aldehydes to the corresponding acid and is part of a novel pathway for carbohydrate oxidation in this organism. The formaldehyde oxidoreductase of *Thermococcus litoralis* converts formaldehyde to formate, but this is thought to be involved in the oxidation of amino acids inside the cell. These two enzymes are unusual in that their catalytic sites contain tungsten, an element seldom found in biological systems.

Dehydrogenase-type enzymes utilize NAD or NADP as electron carriers rather than ferredoxin. However, these cofactors have halflives at 100°C of only a few minutes and it is not known how they are stabilized inside the cell. Interestingly, the half-life of *Thermotoga* lactate dehydrogenase at 90°C increased from 2 minutes to 150 minutes when NAD was present, suggesting that the cofactor and the enzyme may stabilize each other at the growth temperature of the organism.

Table 2.3 : Properties of Enzymes Purified from S°-Reducing Hyperthermophiles

Enzyme	$t_{50\%}$ (h/°C)	Catalytic Activity
Pyrococcus furiosus		
Protease	33/98	Peptide hydrolysis
Amylase	2/120	Starch hydrolysis
α-Glucosidase	48/98	Maltose hydrolysis
Sucrose α-glucohydrolase	48/95	Sucrose hydrolysis
Hydrogenase	2/100	H_2 production (Fd)
Ferredoxin	>24/95	Electron transfer
Rubredoxin	>24/95	Electron transfer
Aldehyde oxidoreductase	6/80	Aldehydes to acids (Fd)
Pyruvate oxidoreductase	0.3/90	Pyruvate to acetyl CoA (Fd)
Glutamate dehydrogenase	10/100	Glutamate to 2-OG (NAD/P)
DNA polymerase	20/95	DNA replication
Thermococcus litoralis		
Ferredoxin	>24/95	Electron transfer
Formaldehyde oxidoreductase	2/80	Formaldehyde to formate (Fd)
DNA polymerase	7/95	DNA replication
Pyrococcus woesei		
Amylase	6/100	Starch hydrolysis
GAPDH	0.7/100	GAP to BPG (NAD/P)
Thermoproteus tenax		
GAPDH	0.3/100	GAP to BPG (NAD)
GAPDH	0.5/100	GAP to BPG (NADP)
ES-4		
Amylopullulanase	20/98	Starch hydrolysis
Pyrodictium brockii		
Hydrogenase	1/98	H_2 oxidation (unknown)
Desulfurococcus mucosus		
Protease	1.5/95	Peptide hydrolysis
Thermotoga maritima		
4-α-Glucanotransferase	3/80	Starch hydrolysis

Hydrogenase	1/90	H_2 production (unknown)
GAPDH	2/95	GAP to BPG
Lactate dehydrogenase	1.5/90	Pyruvate to lactate (NAD)
***Thermotoga* sp. strain FjSS3-B.1**		
Xylanase	1.5/95	Xylan hydrolysis
Cellobiohydrolase	1.1/108	Cellulose hydrolysis

Similarly, the substrates for glyceraldehyde 3-phosphate dehydrogenase (GAPDH) are very unstable, and accurate assays are not possible much above 70°C.

High temperature DNA polymerases are of particular interest because of their use in the polymerase chain reaction (PCR), now a routine tool in molecular biology. Three DNA polymerases are commercially available, and the properties of those from *T. litoralis* and *Ps furiosus* have been described. A thermostable DNA polymerase is essential in PCR work.

STRUCTURAL INFORMATION ON HIGH TEMPERATURE ENZYMES

So far there is only limited information on the enzymes and proteins that have been purified and characterised from hyperthermophilic S°-reducing organisms. From the Archaea, five complete amino acid sequences have been reported: three derived from proteins (ferredoxin and rubredoxin from *P. furiosus,* and *T. litoralis* ferredoxin) and two from gene sequences. In addition, four complete amino acid sequences have been reported for *Thermotoga maritima* proteins: for lactate dehydrogenase (from the protein), GAPDH, glutamine synthetase, and EF-Tu (from gene sequences). So far the genes for three hyperthermophilic proteins have been expressed in *E. coli (P. woesei* GAPDH, and glutamine synthetase and EF-Tu from *Thermotoga maritima)*. The three-dimensional structure of one hyperthermophilic protein has been determined, namely *P. furiosus* rubredoxin.

3

ENZYME ASSAYS

The assay of enzyme activity is one of the most frequently performed experimental procedures in biochemistry. Enzyme assays are essential for assessing the activity of genetically engineered enzymes, for determining kinetic parameters, for estimating the amount of enzyme present in a cell or tissue, and for following the progress of an enzyme purification procedure. This entry describes the principles of designing and carrying out enzyme assays, but it is not a compilation of assays for individual enzymes. Such information may be found in appropriate volumes of *Methods in Enzymology* and in *Methods of Enzymatic Analysis*.

UNITS OF ENZYME ACTIVITY

The most frequently used definition of enzyme activity is the *International Unit* (IU; also enzyme unit), defined by the Enzyme Commission of the International Union of Biochemistry as the amount of enzyme that catalyzes the transformation of substrate to product at a rate of *one micromole per minute*. It should be noted that the enzyme unit refers to an *amount* of substrate transformed, and not to *a concentration*. Since the rate of change in an enzyme assay is often measured as a concentration change, it is important to take into account the volume of the assay mixture when interconverting measured rates and enzyme units.

The introduction of SI units brought about the proposal that the katal be used as the unit of enzyme activity. The *katal* is defined as the amount of enzyme catalyzing the conversion of substrate to product at a rate of *one mole per second*. This is an inconveniently large unit in terms of amounts of enzyme activity actually measured *in vitro* (or indeed occurring *in vivo* in any single organism) and is less often used. The conversion factors for international units and katals are as follows:

$$1 \text{ Kat} = 6 \times 10^7 \text{ IU}$$
$$1 \text{ IU} = 16.67 \text{ nKat}$$

Specific activity, a useful term for describing the purity and activity of an enzyme, is usually expressed as *enzyme units per milligram of protein.* The *molecular activity* of an enzyme is defined as *enzyme units per micromole of enzyme.* If the molecular weight of the enzyme and the purity of the preparation are known, the molecular activity may be calculated from the specific activity. Molecular activity is expressed in units of reciprocal time, and is equivalent to the catalytic constant, k_{ca}. It is the number of molecules of substrate transformed to product per molecule of enzyme per unit time (conventionally per second), and is a measure of the catalytic power of the enzyme. *Turnover number,* a term still occasionally used, is equivalent to the molecular activity, but is sometimes used to refer to the catalytic site activity, which is the molecular activity divided by the number of catalytic sites per enzyme molecule.

When reporting the results of activity assays, information on the pH and composition (including ionic strength) of the assay buffer, and the temperature, should always be stated, as all these affect enzyme activity.

INITIAL RATES

It is essential for proper design of an enzyme assay that the measured activity be proportional to the total concentration of enzyme present. For this to be true, the *initial rate (i.e.,* the steady state rate) must be determined. Most enzyme assays are conducted at near-ambient temperatures, with the initial substrate concentration S_0 greatly exceeding the total enzyme concentration E_0. The condition $S_0 >> E_0$ ensures the validity of the steady state assumption and enables the rate of substrate disappearance to be equated with the rate of product appearance, so that either may be followed to assess the rate of the reaction. Under these conditions, the steady state is usually attained well within one second of mixing the enzyme with the substrate(s). From this point on, the rate (the change of product concentration per unit time) will usually decrease. Provided that an accurate estimate of the rate can be made at "zero" time, the rate may be termed an initial rate and will be proportional to the enzyme concentration. Another reason for using the initial rate is that S_0 is most accurately known (at zero time). It is usually desirable to keep $S_o > 10 \times K_m$ (but beware substrate inhibition, see Section 3.1). This condition not only makes the rate relatively insensitive to errors in S_o but also increases the time over which the initial rate maintains near-constancy. If the progressive decrease in rate as the reaction proceeds occurs to a sufficient extent to make the estimate of initial rate inaccurate, the measured rate will not be proportional to the enzyme concentration. For this reason, continuous assays are to be preferred to discontinuous assays, since the former, deviations of the progress curve from linearity will be easy to identify. Suitable methods may then be applied to correct for the curvature, or the assay method itself may be altered to improve the linearity of the progress curve.

WHY DOES THE RATE DECREASE?

The most common causes for downward curvature in the reaction time course (or *progress curve)* are following described.

Substrate Depletion

Unless a method exists for maintaining the substrate concentration at a constant value (e.g., by recycling the product), substrate depletion is an inevitable consequence of the progress of the reaction. For this reason the initial substrate concentration is usually set at a value greater than K_m, the Michaelis constant of the enzyme (ideally $= 10 \times K_m$), so that the rate will change little as substrate is used up. However one should always take care not to use unnecessarily high initial substrate concentrations, which could lead to substrate inhibition, a not uncommon phenomenon. This in turn could give rise to progress curves in which the rate actually increases as the substrate is used up. High substrate concentrations may also result in the depletion of cofactors such as metal ions that may be essential for enzyme activity.

The effect of substrate depletion may be minimised by observing the rate of reaction over a period of time such that the fraction of the total substrate concentration consumed is very small. This may be accomplished by increasing the sensitivity of the method of detection to allow the determination of reaction rate over a shorter time period or by decreasing the amount of enzyme in the assay. A serious case of substrate depletion results if the substrate is chemically unstable under the assay conditions. Although it may be possible to correct for such instability by analyzing the kinetics of the decomposition reaction, it is preferable to alter the assay conditions so that the substrate is stable. For example, assays involving

NAD(P)$^+$-dependent dehydrogenases are rarely carried out at pH values much below 7 because NAD(P)H is unstable under acid conditions.

Approach to Equilibrium

If the catalyzed reaction is reversible, the contribution of the backreaction to the net rate will increase as the reaction approaches equilibrium. This effect can be minimised by running the assay under conditions that shift the equilibrium in favor of product formation. Thus, enzymes that catalyze reactions producing hydrogen ions, such as those involving the conversion of NAD(P)$^+$ to NAD(P)H, are often assayed at pH > 9.

A related approach is to use trapping reagents that remove the product as it is formed. An example of this is found in the assay of alcohol dehydrogenase, which catalyzes the oxidation of ethanol by NAD'. This assay is often run in the presence of semicarbazide, which reacts with the acetaldehyde produced to form the semicarbazone and thus removes the product from the reaction mixture. If the purpose of the assay is simply to assess the amount of enzyme activity, it may be helpful to run the assay in reverse [i.e., to interchange substrate(s) and product(s)]. For example, at pH values below 9, the equilibrium of the reaction catalyzed by alcohol dehydrogenase favors the formation of alcohol; thus the progress curve is linear for longer periods of time when assays are conducted in this direction.

Product Inhibition

Curvature of the progress curve can result from product inhibition even when the reaction is effectively irreversible. This effect is sometimes so severe that it is necessary to sacrifice the convenience of a continuous assay for the accuracy of a more sensitive discontinuous one. The only way to eliminate this phenomenon (i.e., product inhibition) is to remove the product from the assay as rapidly as it is formed. In the assay of catechol-O-methyltransferase (COMT), the enzyme is powerfully inhibited by the reaction product, Sadenosylhomocysteine. This can be removed by inclusion in the assay of adenosine deaminase (ADA), which converts the inhibitory product into S-inosylhomocysteine, which does not inhibit COMT.

Inactivation of Enzyme

Inactivation of the enzyme during the assay will also cause a decrease in rate and curvature of the time course. However, the remedies suggested in the chapter elsewhere are unlikely to remove the cause. Indeed some of the measures suggested (e.g., dilution of the enzyme) may well exacerbate the problem if, for example, the cause of the inactivation is irreversible dissociation of the enzyme into inactive subunits, or adsorption on glass. A simple graphical test for enzyme inactivation during an assay is available and should be applied whenever this is suspected. If inactivation is found to be occurring, one or several approaches may be tried.

Proteins are frequently stabilized by the presence of other proteins; for this reason serum albumin is often included (at 0.1-1.0 mg/mL) in assay mixtures. If the cause of enzyme inactivation is oxidation of -SH groups, addition of thiol reagents such as dithiothreitol or mercaptoethanol may help, although in extreme cases it may be necessary to use degassed solutions and to exclude oxygen from the assay. Trace heavy metals in assay components may also cause loss of enzyme activity; for this reason chelating agents, such as EDTA, are often included in assay mixtures. The presence of glycerol (5-20% v/v) has also been reported to stabilize enzymes in dilute solution. Treatment of glass vessels with siliconizing agents or the use of plastic vessels may help if inactivation is due to adsorption.

Artifactual Cause

A well-designed assay should not be plagued by artifacts, but one should be on guard for the presence of, for example, inadequate buffering in a reaction involving production or consumption of hydrogen ions, inadequate temperature control of the assay or addition of cold assay components, and nonlinearity of the instrumental method (frequently spectrophotometric) used for the assay.

TYPES OF ASSAYS

Assays can be continuous or discontinuous, direct or indirect.

Continuous or Discontinous Assay?

Continuous assays may be used whenever a sufficient difference exists between a directly detectable property of a substrate and product. Most commonly, continuous assays utilize changes that may be followed photometrically, fluorimetrically, polarographically, polarimetrically, or through changes in pH that occur as the reaction proceeds. A great advantage of such assays is that the shape of the progress curve is revealed, and any irregularities in curvature are readily detected. Such information not only simplifies initial rate estimation, but where anomalous behavior is observed, it may disclose interesting properties of the enzyme. A possible disadvantage is that continuous assays must be carried out sequentially. Automated instrumentation is commercially available and may be useful where many samples must be processed under identical assay conditions. For example, enzyme-linked immunosorbent assay (ELISA) plate readers, which can read 96 samples almost simultaneously and repeatedly, enable a large number of continuous assays to be processed in one operation.

Discontinuous assays, sometimes termed stopped assays or sampling assays, are carried out by terminating the reaction by inactivating or removing the enzyme and/or stopping the reaction, often by raising or lowering the pH of the assay mixture or of a sample withdrawn from the assay. Substrate and product are then separated, or the quenched reaction mixture is treated to produce a chemical change in either substrate or product (e.g., a colour reaction), allowing one or the other to be detected. Discontinuous assays have two disadvantages compared with continuous ones. First, the shape of the progress curve is not readily apparent (unless samples are taken at many time points). Also, timing and volume inaccuracies may be introduced, associated with the termination of the reaction and withdrawal of samples at fixed times. Thus it is especially important to ensure that rates estimated using such assays are truly proportional to enzyme concentration. However, discontinuous methods do allow many assays to be run at the same time. They are also inherent in the use of sensitive and selective assay techniques that require separation of product and substrate either prior to (as in radioactivity measurement) or as an integral part [as in high performance liquid chromatographic (HPLC) techniques] of the measurement.

Direct Assays

The spectrophotometric assay of xanthine oxidase is a good example of a direct continuous assay. Here the substrate, xanthine, is converted into uric acid. Both substrate and product absorb light in the UV region but have different absorbance maxima. The difference between the absorption coefficients of substrate and product is greatest at 292 nm-a convenient wavelength to use for the assay. Many HPLC and radiochemical assays are examples of direct discontinuous methods.

Indirect Assays

Many assays are indirect, in that the change in product or substrate concentration is measured after a subsequent chemical reaction has produced an observable phenomenon (e.g., a colour change). Coupled assays are an important example of indirect assays.

Coupled Assays

Although it is usually wise to keep experimental protocols as simple as possible, coupled enzyme assays are often used. In this procedure, a reaction of interest is linked to one or more subsequent reactions. Coupled assays may be continuous or discontinuous. As an example of the former, one may use the often-employed pyruvate kinase-lactate dehydrogenase (PK-LDH) system. This enzyme couple is used to assay enzyme-catalyzed reactions that produce ADP [e.g., glycerol kinase (GK)]:

$$\text{glycerol} + \text{ATP} \xrightarrow{\text{GK}} \text{glycerophosphate} + \text{ADP}$$

$$\text{ADP} + \text{phosphoenolpyruvate} \xrightarrow{\text{PK}} \text{ATP} + \text{pyruvate}$$

$$\text{pyruvate} + \text{NADH} \xrightarrow{\text{LDH}} \text{lacate} + \text{NAD}^+$$

This system allows GK to be assayed continuously by observing the decrease in absorbance at 340 nm as NADH is converted to NAD+. For this, the assay mixture would contain phosphoenolpyruvate, PK, LDH, and NADH, in addition to GK, glycerol, and ATP. The disadvantages of increasing the complexity of the assay are outweighed by the simplicity of the continuous spectrophotometric method. As a bonus, the substrate, ATP, is recycled. An example of a discontinuous coupled assay is that described for COMT. Here the disadvantage of introducing an additional component (ADA) is more than compensated by the removal from the system of the strongly inhibiting product.

In a coupled assay, the rate actually measured should equal the rate of the reaction being catalyzed by the enzyme under investigation. Thus it is essential that the coupling enzyme(s) be present in sufficient excess (in activity terms) over the enzyme under study to prevent the rate of the coupling reactions from becoming limiting. There is also an inevitable lag in such assays as the substrate(s) of the coupling reaction(s) build up to their steady state levels in the reaction mixture. Increasing the amount of coupling enzyme in the assay will reduce this lag. However, it is unwise simply to add a "vast excess" of coupling enzyme. In addition to being wasteful and costly, this practice may lead to unexpected complications if the coupling enzyme contains contaminating enzyme activities that interfere with the overall assay. Methods are available for the calculation of the amounts of coupling enzyme necessary to give desired lag times and ratios of measured to true activity.

ESTIMATION OF INITIAL RATE

If a progress curve is sufficiently linear, the initial rate may be estimated by eye and a ruler. When fitting by eye, care must be taken to avoid subjective bias. For a reasonably accurate estimate, the extent of reaction followed should be no more than 5%, and the substrate. concentration should be no greater than 5 X K_m. In some circumstances (e.g., the equilibrium is highly unfavorable, or the substrate concentration necessarily is less than K_m), it may be impossible to avoid highly curved time courses. In these cases it is best to use an objective analytical method, such as fitting the time course to a polynomial. A somewhat less empirical approach based on the integrated Michaelis-Menten equation is also available. Such methods are equally applicable to discontinuous and continuous assays (provided a sufficient number of points are obtained). However, when many discontinuous assays are to be performed under varying conditions (e.g., of substrate or inhibitor concentrations), determination of a progress curve of several points for each assay may involve an impracticably large number of experiments. In such cases it is usually worthwhile to try to linearize the initial portion of the progress curve by altering the fixed conditions of the system. Such investigations should be carried out under conditions of the variable(s) giving the highest and lowest rates in the experimental series. If the progress curves are linear for these rates, it is usually safe to assume that they will be linear over that time period for all assays measuring intermediate rates. As a further check, proportionality of rate with enzyme concentration should be determined for conditions giving the highest and lowest rates.

Lags or bursts in progress curves, which may cause problems in initial rate estimation, can arise for artifactual reasons (instrumental response, dust particles, temperature equilibration, etc.) or from factors inherent in the assay or the enzyme, such as product activation, pre-steady-state transients, or hysteretic effects. It is obviously essential to determine the cause of such behavior, to ensure that the correct portion of the progress curve is used to estimate the initial rate. Once a reliable method for initial rate estimation has been established for a given set of assay conditions, it is usually safe to assume that it will be valid for all smaller rates measured.

LINEARITY OF MEASURED RATE WITH ENZYME CONCENTRATION

Provided the considerations discussed above have been met, a plot of activity versus enzyme concentration will be a straight line going through the, origin. If the activity is expressed as a molar concentration change, if the substrate concentration is saturating, and if the molar concentration of the enzyme is used, the slope of this line will be the turnover number. However deviations from linearity may occur, and it is important to realize when these are due to some property of the enzyme are a result of an experimental artifact.

Failure to subtract a blank rate will result in a straight line that intersects the ordinate axis, usually above the origin. Blank rates (i.e., nonenzymic rates) may arise from several causes. The most obvious is the occurrence of the uncatalyzed reaction. However instability of the detected substrate, e.g., resulting from the presence of NADH oxidase in the GK assay system, would also result in a blank rate. Whenever assay conditions are altered, such rates should be determined and subtracted from the observed enzymic rate. Any artifactual situation that affects the shape of progress curves, such as instrumental limitations, may also result in deviation from linearity of the relationship of rate to enzyme concentration. It should be realised that such behavior may be a result of a property of the enzyme itself. If the enzyme dissociates into subunits that are inactive, the plot of velocity versus E_0 may show upward curvature, since the fraction of enzyme in the active form will increase with increasing E_0. The converse will obtain if it is the dissociated form is active and the aggregated form that is inactive. Be warned that similar results may be obtained if the enzyme stock solution is contaminated with a reversibly dissociating activator (upward curvature) or inhibitor (downward curvature).

4

TRANSPORT

The term "*transport*" refers to a physical event, not to a particular mechanism or purpose. When we talk of the transport of a substance across a membrane, we are not saying how it crosses, or why, but simply describing a phenomenon. This, as we have already discussed, is different from talking of "the *secretion* of digestive enzyme by *exocytosis*," in which case we are not only saying that protein is transported out of the cell, but that it leaves in a particular way (*exocytosis*) that is purposeful (*secretion*).

This having been said, however, the word "*transport*," as it has been commonly applied to the movement of molecules and atoms across short distances such as biological membranes does imply something about mechanism at the "*microscopic*" level. That is, the movement of the transportable substance occurs as a result of its interactions with nearest neighbor molecules and atoms due to the kinetic energy of the individual molecules themselves. It is these interactions that are thought to account directly for the movement of a particular molecule within a phase or from phase to phase.

The transport of large organic molecules such as proteins has been considered sui generis; that is, as a system apart from this general perception of how molecules move across membranes. This is because such transport has been considered, until very recently, a "*macroscopic*," rather than a "*microscopic*" process. For example, the secretion of macromolecules is thought to occur as the result of the release of hundreds of thousands or even millions of molecules from the cell at once through holes in the plasma membrane produced by exocytosis. In this case, microscopic interactions between individual permeating molecules and the plasma membrane would not occur, and their transport would instead occur en masse as the result of the integrity of the membrane barrier or boundary being compromised temporarily by exocytosis. Characterizing such transport in "*microscopic*" or general transport terms would be inappropriate. Even when models have been considered in which the protein is thought to interact with the membrane directly (binding), as in certain *pinocytotic* or *endocytic models*, transfer across the barrier is still thought to be the result of its removal. Hence, even though interactions between the permeating molecule and the membrane may occur, they do not directly account for the passage of the molecule across it. It is this line of thought that makes it possible to say that membranes are impermeable to proteins and at the same time know that protein transport across them is a process that occurs throughout animal and plant kingdoms.

ELECTRON MICROSCOPY

Placing protein transport as an independent subject has in part been the result of historical circumstances that have led to the process being studied predominantly from an "anatomical" or visual perspective. *Cell biology*, the child of classical anatomy and histology, came into full flower after the Second World War with the development of the electron microscope, which made it possible to look into the cell and describe with some apparent precision intracellular parts and organelles. With the increased power to resolve small objects that this instrument gave us, cells that had merely seemed to be sacs of nuclei-containing protoplasm, were found to be far more complex and visually interesting. Many new features were observed, and structures only guessed at in the *light microscope* could now be confirmed at the ultrastructural level.

Like many of their predecessors, anatomists of ultrastructure were not content to just describe what they observed. They also wanted to know what these objects did. Attempts to answer this question often followed a particular line of thought. If a cell was known to carry out a particular function, then it was reasonable to think that structures found within them might be related to that process in some way or another. This line of reasoning was further validated by the widely held, although often unstated, view that cells of eukaryotes, particularly higher forms, are differentiated to the ultimate degree; that is to say, in their differentiated state virtually all of the cell's activities subserve one particular or special function such as muscle contraction, secretion, etc. If one could indeed assume that contraction was what a muscle cell was up to, then it made good sense to try to relate muscle ultrastructure to this function.

One could seek clues about function from the form and location of particular objects in cells. Considering muscle, as detailed knowledge of the ultrastructure of the striated muscle fiber became known, it was natural to ask how the structures running parallel to the long axis of the cell (thick and thin filaments) might be related to the contractile process that occurred along this axis. Thus, by assuming that a particular form is related to a particular process, and by then trying to guess what its role in that process might be from its shape and location in the cell, it was possible to develop specific models for processes at the ultrastructural level that could then be tested.

Thus, the greatly increased resolution that the electron microscope gave to cell biologists led to the development of a great many *new hypotheses* that attempted to explain how particular cellular functions are carried out by trying to relate observed structures to known functions. This was certainly not a new pattern of hypothesis development, but nevertheless, the increased focus of our vision had led to the development of functional models based on structures whose existence was made known by being able to examine the system at a higher magnification, closer to the molecular level. Thus, experimental reductionism had another success. The technological advance that the *electron microscope* embodied led to a great increase in our knowledge of the cell's structures, and this in turn was food for our imagination as we tried to relate these structures to the known functions of cells.

ATTENDANT PROBLEMS

Even though the visual resolution of forms inside of cells was increased by several orders of magnitude, being able to resolve even certain single molecules, the material that we examined in this way was static and related to the living state in often unclear and ill-defined ways. Fixation, staining, dehydration, and other treatments required for electron microscopy alter samples in ways that we often cannot explicitly predict or characterize. What we see is a generally agreed upon distortion of reality that we hope, with some reason, bears a resemblance to the real thing. Of course, this is true of all scientific theories, but unfortunately when we look at "*pictures*" of "*real*" cells, the fact that the images are merely hypotheses themselves can easily be lost.

But perhaps more to the point here than the accuracy of the images is the fact that they are invariably static. We cannot actually follow continuous (*dynamic*) processes with this technique. The scientist looks at his or her material and from it imagines how the images may be related to actual occurrences in the living system. At best we can examine samples of material taken at different points in time and then attempt to "*reconstruct*" actual occurrences as we think they might occur in real space-time. It is sometimes thought that this is wholly analogous to following cytologic changes in living cells. Unfortunately, it is not, unless of course we have independent evidence that such changes occur in living cells in the first place. If we have an hypothesis that a cytologic change of a certain type occurs and observe electron microscopic images that are consistent with this hypothesis, then we have provided evidence in its support (weak inference), but have not described the events involved in the natural process independent of our hypothetical construction. This approach to the discovery of physiological mechanisms when used in isolation has the important limitations.

THE IMPORTANCE OF SPECIFIC MECHANISTC HYPOTHESES

The term "*transport*" is descriptive of a phenomenon, and the study of transport in biology is in the first instance the delineation by direct measurement of the *kinetic* and *thermodynamic* characteristics of a particular transport process. How fast does transport occur? How is rate related to the concentration of the transported substrate? How do alterations in the structure of the permeating molecule alter its rate of transport? And so on. It is on the basis of these measurements that we can then build models that attempt to incorporate the observed properties of the system and will be useful in predicting its responses. These models may also be used to characterize the underlying mechanisms in specific molecular terms. Only with such knowledge of the system's properties in hand, can we have confidence that the isolation (*chemical* or *anatomical*) of what we think are parts of the system indeed relate to the natural processes in a way similar to what we imagine. That is, we can only have this assurance, if we know how the system should behave however it is put together; i.e., if our initial description of the system is accurate and broad. In this case, our "*reductionist*" or "*atomistic*" model can be compared with the actual behavior of an intact system. Thus, our approach to the study of molecular transport has at its center the actual physico-chemical description of the system from which reductionist or atomistic science attempts to determine its detailed structure.

Thus, in examining a transport process we must first describe the actual or "*real world*" properties of the system before developing detailed models at the molecular or other "*reducing*" level. To do so in the absence of this prior knowledge in a sense presumes that the investigator has devised an hypothesis that is adequate and that other possibilities have not escaped attention; in essence, we assume that the hypothesis is correct, at least in general terms, before we even test it. In this case, our hypothesis and the description of the system may become confused with each other. This has occurred in the study of protein transport wherein the vesicle hypothesis and what is thought to be a description of the system have in some instances become virtually indistinguishable.

STRUCTURE-FUNCTION HYPOTHESES

The presence of particular visual images within cells and knowledge that a certain function is carried out by a cell permits us to propose or assume a connection between the two; but it, of course, does not demonstrate one. In some cases, as with muscle contraction,

1. The discovery of a process or event.
2. A description or characterisation of that process or event in real space-time (kinetic, thermodynamic, etc).
3. Attempts to identify and isolate the parts of the system.
4. Reconstruction or other tests to demonstrate whether the isolated parts can be reassembled or otherwise can account for the behavior of the intact system in terms of a particular model.

Figure 2.1 : A sequence for the investigation of a biological process or event.

considerable knowledge of the function (or *physiology*) of the intact process from a dynamic and thermodynamic point of view existed prior to electron microscopy studies that attempted to correlate structure and function. Therefore, a model constructed from forms seen within the cell could be based on and tested against substantial knowledge of the system's actual behavior. Whatever limitations this approach may have for muscle contraction, attempts to make similar inferences about the relationship between the structure and functions of the acinar cell of the pancreas has been far more problematic.

CELLULAR PERMEABILITY

When the *electron microscope* permitted the first detailed look into the world of the cell, the acinar cell of the pancreas was particularly interesting. As with striated muscle, it was also quite natural to wonder how the forms seen in the microscope might be related to the cell's ability to perform a particular function, namely, the secretion of large quantities of protein, and to then build a model for secretion based on these forms and their location in the cell.

The most striking structures were the spherical zymogen granules that were visible in the *light microscope*. But in addition, one could now see what the Golgi stain did stain: discontinuous stacks of membrane-bound cisterns, the *Golgi complex*. And the staining characteristics of the basal portion of the cell could be attributed to small, roughly spherical structures, *ribosomes*, that we now know are involved in *protein synthesis*. Most of the ribosomes appeared to be attached to numerous parallel flattened membrane saccules, the rough-surfaced endoplasmic reticulum or RER (called rough due to the attachment of the ribosomes to the membrane's surface).

A microscopist looking at this bewildering array of subcellular detail might well attempt to develop an hypothesis for protein secretion that would relate these structures to each other and to protein synthesis, intracellular transport, storage, and the release of digestive enzyme proteins from the cell. *Zymogen granules*, predominantly located at the cell apex, contained digestive enzymes and their disappearance had been associated with secretion in the past. Furthermore, it was thought that proteins were synthesized in association with ribosomes that were localised primarily in the basal or blood-facing portion of the cell. Thus, new protein had to get from the ribosomes at the cell's base to the zymogen granules at its apex, and from there into the duct lumen. But how did this occur? How might the objects seen under the microscope be involved in the movement of this material? What led to the disappearance of granules? What was the role of the *RER* and the *Golgi complex*, assuming that they had one? The pattern of hypothesis development flowed from questions of this nature.

The task then appeared to be to relate structures to events and to overcome the limitations of inferences based solely on static electron microscopic images. But against what information could a putative structurefunction model be tested? Unlike the muscle fiber, there was not much in the way of kinetic or thermodynamic knowledge about the intact secretory system. In muscle, our "*reductionist*" or "*atomistic*" hypothesis must in the end account for the knowns of contraction: i.e., its physiology, its various heats, the length-tension relationship, the timecourse of the process, its latency, etc. Although this fit may have yet to be satisfactorily achieved, what is required of a successful model is relatively clear.

Unfortunately, a similar body of knowledge against which a model could be tested did not exist for protein secretion by the acinar cell. Indeed, little was actually known about the transport process beyond the fact that the enzymes were secreted into the extracellular space (the duct system), that a variety of different proteins were secreted, that the concentration of protein in secretion was variable, and that the zymogen granules that contained these proteins appeared to disappear when secretion was stimulated.

Thus, a model that attempted to describe the mechanism of protein secretion could not be much more than a construct from the imagination, based on assumptions about the system; a structural model for a physiological process evoked by static images in the absence of much information about the kinetics and thermodynamics of the intact process that it hoped to explain.

VESICLE FORMATION

Thus, the *vesicle* or *cisternal packaging—exocytosis* model for protein secretion by the pancreatic *acinar cell* is an example of a constructed hypothesis; one built to a considerable degree on a series of assumptions about the system and knowledge of its presumed parts in the relative absence of a foundation of accurate knowledge of the intact system whose behavior it seeks to describe. The knowledge of secretion that we have gained since its proposal more than 25 years ago has for the most part been limited to information gathered by testing, by weak inductive inference, one aspect or another of this particular model.

The model can be summarised briefly as follows. New protein chains are "*sequestered,*" or *segregated*, within the cisterns of the *endoplasmic reticulum* as they are being synthesized on *ribosomes* attached to the membrane surface of this structure. The secretory products contained within RER cisternae are then moved to the *Golgi* region of the cell in small vesicles thought to bud from the RER and subsequently fuse either with *Golgi membranes* or the membranes of filling zymogen granules, called *condensing vacuoles*, wherein the small vesicles release their contents. Secretion granules fill in this way and eventually form mature zymogen granules that are capable of fusing with the plasma membrane of the cell in the exocytosis event leading to the final and irreversible release or secretion of product. Palade and Jamieson claim that the following is known or established:

"*Transport* along the secretory pathway is: (a) *vectorial*, i.e., it proceeds regularly in the direction of secretion granules (apically) in exocrine cells; (b) *irreversible*, i.e., there is no evidence of back flow; and (c) *functionally discontinuous*, i.e., the secretory products are moved from one compartment to the next in series by discrete membrane containers (*vesicles* or *vacuoles*) and the pathway behaves as if provided with locks whose opening requires energy and involves membrane fusion-fission."

ASSUMPTIONS

However substantial or weak, clear or ambiguous, one considers the evidence that has been offered over the years in support of the paradigm, the issue rested in the minds of many, and perhaps still does, on our assumptions about the system. If we can assume that a mechanism similar to the one that has been proposed must exist, then whatever weaknesses the evidence may have as convincing proof dissolves into insignificance, because we really have little choice in the matter.

The *vesicle hypothesis* has in great part been based on, and has its origins in, two particular assumptions. If we accept either of them, then we have a priori excluded the only apparent general alternative to it: a diffusive or diffusion-based transport process in which proteins move individually, interact directly with and move through membrane elements in order to effect transport. That is, if we believe these assumptions to be necessary and sufficient, then the existence of a vesicular system in which proteins are moved en masse is established without the need for direct evidence of its existence at all.

The First Assumption

The *major assumption* is that proteins cannot cross biological membranes individually; that is, biological membranes present an absolute barrier to this class of molecules, and therefore other means must exist to account for their known movement across them. This assumption was thought to be quite compelling during the period in which the paradigm was being developed.

The cisternal packaging-*exocytosis* paradigm was itself based on another paradigm, the then-current hypothesis for the structure of biological membranes, the *Davson-Danielli model*, which proposed that biological membranes are comprised of a continuous lipid bilayer to which proteins are appended at the surface, binding to the polar groups of the lipids, particularly phospholipids, facing the two aqueous environments (cellular and extracellular fluid) bordering the bilayer. Proteins (at least globular and so-called random coil proteins) were seen as having considerable, and essentially invariant, surface charge that prohibited their entrance into the resolutely nonpolar center of the membrane. Hence, charged macromolecules could not enter a membrane, no less cross it. This view of impenetrability was further supported by the fact that many secretory proteins were water-soluble globular proteins that were insoluble in many bulk nonpolar solvents. Thus, the idea that membranes were impermeable to proteins, left us to find other means to explain their known movement across them, hence vesicle mechanisms.

SEMIPERMEABILITY

Our ideas about biological membranes and their permeability are in great part derived from the concept of "*semipermeability*" first introduced by Nageli in the middle of the 19th century. Semipermeability simply meant that membranes were permeable to some substances, water in particular, but not others. In this regard, the cell membrane has been thought to have two barrier functions. The first is to *permit,* or even facilitate, the entrance of needed substances, such as water and nutrients, to which it is permeable. The second is to *prevent* the entrance of other (harmful) substances or the loss of important cellular components such as proteins and nucleic acids, to which it is impermeable.

At first biological membranes were thought to be impermeable to most substances. For many years, even the sodium ion, whose entrance into cells is now not only well established, but considered crucial to cellular function, was not thought to enter cells. At first indirectly, and then most notably with the advent of isotope tracer techniques, it was shown directly that even though the rate of sodium permeation was rather slow compared with some other atoms and molecules, it did enter cells at a measurable rate, and that over time the cell would, if permitted (in the absence of countervailing processes), come into electrochemical equilibrium with external sodium.

In fact, as experimental data accumulated it became clear that cells were permeable to a very wide range of substances, even ones that they might never see in the normal course of events (such as those made in the laboratory), or that might be harmful to them (such as urea is to red blood cells, destroying the cell just because it enters).

Thus, the *permeability* of cells was a general phenomenon, and whether or not a cell was permeable to a particular substance did not necessarily reflect either its usefulness or its potential danger. Whether a molecule did or did not cross the barrier simply depended upon its physical and chemical characteristics in relation to those of the membrane.

It was also realised that the concentration of a substance in a cell was determined by the *algebraic sum* of its rate of entrance and exit. That is, substances did not merely enter or exit, but, as with *simple diffusion*, permeability was bidirectional, and intracellular concentration was determined by the relative rates of both influx and efflux (or outflux). In addition, for nutrients such as sugars and amino acids and products of cellular metabolism such as organic acids, cellular concentration was also determined by a third factor, the rate of utilisation or production. In this case, the observed intracellular concentration of course reflected a balance between influx, efflux, and rate of utilisation or production.

Despite the fact that the sodium ion enters cells, its concentration in many cells, notably those that have been widely studied, such as the red blood cell, nerve, and striated muscle, is very low in comparison to extracellular biological fluids (N.B.: A central reason why the membrane was thought impermeable to it). This presented a *paradox*. How could sodium both enter the cell from a high-concentration external medium and yet remain at a low concentration within the cell at the steady state? This problem was resolved when it was demonstrated that the cell could maintain a dynamic disequilibrium in which the rate of sodium efflux for a given concentration was greater than its rate of entrance; hence, the ability to maintain a low internal concentration at the steady state. That is, the cell's sodium content constantly changes due to influx and efflux, but nevertheless a low concentration is maintained by differences in the respective transport "*affinities*" (permeabilities) that makes it "*easier*" for sodium to exit than enter the cell. Thus, it became clear that the cell had "*special*" processes that could produce and maintain nonequilibrium steady or stationary states across the membrane for sodium, and, as it was discovered, numerous other molecules as well.

In this way, our view of the presence or absence of particular substances in cells has gradually evolved from one in which the plasma membrane was seen as being permeable to some molecules, but not others, to one in which the presence or absence of a substance in a cell (or whether it is present at a high or low concentration) in itself tells us nothing about whether or not it crosses the cell membrane or if it does, at what rate. The idea that the cell's contents of a molecule reflects a distribution between cell and environment in which the observed steady-state concentration is determined by a continuous, dynamic balance between influx, efflux, and utilisation or production when that occurs has become the paradigm for the great majority of biologically significant substances.

Nevertheless, despite the realisation that the membrane is permeable to a wide range of substances, that this permeability is bidirectional, that the degree of permeation varies broadly for different molecules and membranes, and that nonequilibrium steady states can be maintained in biological systems, the dichotomous view that a membrane is either permeable to a substance or impermeable to it has remained with us over the years. However the list of substances thought to be absolutely excluded has gradually decreased as our knowledge has grown. The most prominant members of this dwindling list of excluded molecules are macromolecules, particularly proteins and nucleic acids. Thus, although the older view of absolute impermeability has gradually changed to a more relativistic, statistical, and dynamic conception, it has not changed for proteins. Although as we shall see later, it has been nibbled away at the edges. And even though it is clear that intracellular proteins, for example, exist in a dynamic steady state, their concentration is thought to be solely due to a balance between the rate of their manufacture (synthesis) and subsequent breakdown (by enzymatic degradation) within a cell-the possibility of their exit or entrance is generally excluded a priori.

MEMBRANOUS PORES

Some membranes were found to be *permeable* to many water-soluble molecules, including water itself, at rates that were seemingly substantially greater than predicted for their transport across a *lipid bilayer* by solution and diffusion. This posed another paradox. If the membrane is formed by a continuous lipid bilayer, then why should the rate of transport be greater than predicted for diffusion through such a medium, at least for molecules for which no special mechanism appeared to exist to accelerate or facilitate movement, such as was apparently the case for water? In an attempt to resolve this problem, it was proposed that the membrane was perforated by a mosaic of water-filled pores or channels that transected the bilayer and through which water and other water-soluble substances passed.

Might not proteins exit or enter cells through such channels as well?· The conventional wisdom has been "*no.*" When pore size has been estimated from the observed rates of passage of test molecules across specific membranes, it has frequently (although not always) been found to be significantly smaller (4-10 Å radius) than the radius of most proteins (about 10-30 Å for most globular protein monomers), and thus proteins have been thought to be excluded.

In a sense this was reassuring. If the pores were large enough to accommodate proteins, then how could the cell maintain its integrity? Wouldn't crucial intracellular proteins be lost? Indeed, wouldn't the existence of large pores in the cell membrane simply be incompatible with life? Isn't our concept of the cell membrane as a semipermeable barrier in part that it maintains the life of the cell by preventing the loss of many crucial substances, particularly nucleic acids and intracellular proteins, and in this way helps keep the entropic wolf from the cell's door? Isn't the view that the cell is impermeable to proteins embedded in no less than our understanding of the purpose of the cell membrane, and in a sense even our definition of the cell itself? It goes beyond models of membrane structure.

Regardless of our beliefs about the nature of cells, the purpose of their limiting membranes, or our models of membrane structure, how justified is the assumption that proteins cannot pass through channels or pores, if such they be, or pass through the cell membrane in other ways? Might the factors that determine the steady-state concentration of cellular proteins include a transport *component* as well? It will be considered now, first in terms of the pore concept and then in more general terms.

The permeability of a membrane to a series of water-soluble test molecules drops sharply as the size of the test molecule increases, following functions that appear analogous to the *Stokes-Einstein equation* in which the *diffusion coefficient*, *D*, or the ease with which a substance moves through a medium, decreases as the radius of the particle increases. This analogy is thought to fail however as we consider molecules of increasing size such as moderately sized organic compounds, or biological *macromolecules* such as proteins. In diffusion, although the rate of a particle's movement decreases as its size increases, in dilute solution at temperatures in the ambient range, the limit of zero mobility is not approached, and specifically for molecules of the size of biological macromolecules, mobility is substantial in solution (in water), even though it may be magnitudinally less than for smaller molecules. The *permeability* (P) of biological membranes, on the other hand, is thought to reach a particular discrete limit at which the "*mobility*," or *permeability*, of molecules of greater than a certain specified size can be said to be zero. That is, the movement of molecules larger than the size of the theoretical pore spanning the membrane would be absolutely, and not merely relatively, restricted.

This view is derived in great part from studies that estimate a small pore size by measuring permeability decreases with increasing molecular radius, weight, or volume for a series of chemically *homologous compounds*. However, such measurements do not establish absolute limits for pore size. This conclusion is an extrapolation from the data; it is not demonstrated by it. Indeed, pore size estimates are statistical averages obtained by a combination of experimental and mathematical techniques that permit the isolation of a value for average pore radius from the area component of the overall proportionality constant that relates solute flux across the membrane to concentration difference. Because the value is an average, the conclusion that the calculated pore radius is, for example, 5 Å, does not permit us to conclude that all pores are 5 Å or that 5 Å is the maximum size of a pore. If we assume for convenience that the distribution of pore size is normal and that the standard deviation around the mean value of 5 Å is a modest ± 100%, then on a purely statistical basis we would estimate that 0.5% of the pores would be four times the average

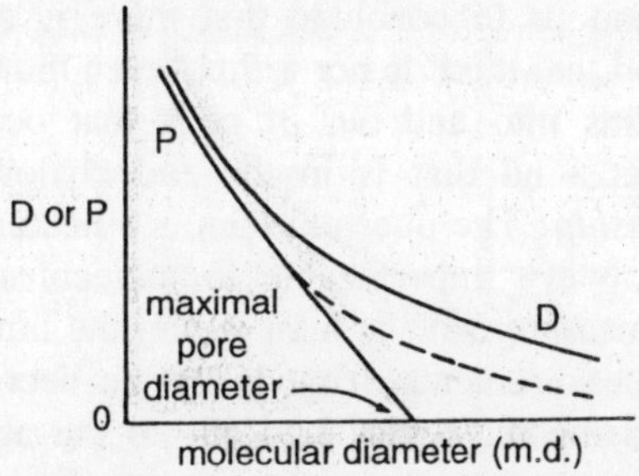

Figure 4.1 : The relationship between the free diffusion coefficient (D) or the membrane permeability coefficient (P) and the diameter of a test molecule (m.d.).

radius, or 20 Å. A pore of this size would be able to accommodate many protein monomers. Thus, estimates of pore radius do not permit us to conclude that there is an absolute limit to the size of permeating molecules. Such a limit may or may not exist, but we can not set it at less than the total surface area of the cell itself from such measurements. We can however say that if the standard deviation of pore size around the mean is rather small, then large pores, if they exist, would be relatively few in number and would only account for a trivial percentage of membrane conductance and total pore area. That is, if such pores or analogous structures exist, and proteins can pass through them, then we would expect the rate of protein passage across the membrane to be very much less than that for small molecules.

But what is the importance of such a conclusion? For example, if we ignore *potential differences* (for example, due to diffusivity or electric field) in the rate of passage through a small (5 Å radius) pore that can accommodate a sodium ion (23 daltons) or a large pore (20 Å radius) that can accommodate a particular 25,000-dalton protein, then for a given concentration gradient, 1000 small pores would move the same mass as one large pore, even though they would nevertheless account for over 99% of the total *pore area*. For the acinar cell, which moves large quantities of protein across the relatively small surface area of the cell's apical membrane, one can calculate that no more than about 0.001-0.01 of 1% (10^{-a}-10^{-5}) of the surface area need be occupied by pores large enough to accommodate proteins in order to account, not only for modest rates of protein secretion, but the maximal observed rate (N.B.: Based on the assumptions that free diffusion through the pores applies and that only one protein molecule can occupy it at a time).

Thus, although there is no disputing the fact that membranes are less permeable to large molecules than small ones, this does not necessarily permit us to conclude that there is an absolute limit to permeability, and that there is not a flux, even though small in relative terms, of proteins into and out of cells that occurs in the normal course of events a nd that is in the end analogous to the *Stokes-Einstein relationship*. The question then is whether or not membranes are indeed absolutely impermeable to molecules of greater than a certain size by measurement, at least within the limits of our capability to make such measurements. That is, is the decreasing permeability seen with increasing molecular size due to the absolute exclusion of molecules of greater than a certain size (is the *reflection coefficient* actually 1.0), or is the exclusion merely relative (the reflection coefficient approaches but does not reach 1.0) ? Since the existence of pores has primarily been inferred from measurements of the permeability of membranes to certain molecules, it certainly seems reasonable to place limits on its permeability (to proteins) in the same way by attempting to measure it directly.

In this regard when a protein is said to "*leak*" from or into a cell, it is often assumed that this is because the cell membrane has been damaged or compromised in some fashion or another that permits the release or entrance of the protein. Whether or not this is the case in one particular circumstance or another, this inference is often drawn with no evidence that the cell membrane is damaged at all, other than its permeability to the macromolecule itself. Of course this is circular reasoning, and is based on the belief that the membrane is normally absolutely impermeable to such molecules. In this case any exhibited permeability must perforce reflect damage and nothing more. Although it certainly is true that if one "*damages*" the cell membrane, at the limit by removing it, that most of the contained molecules will diffuse or otherwise move away; and while it is also clear that an unnatural permeability of membranes to proteins can occur, for example in the release of hemoglobin from red blood cells during hemolysis, we cannot assume that whenever a protein enters or exits a cell (other than by vesicular means) that it is unnatural. This must be proven by evidence in each case, and cannot be assumed a priori, any more than we can assume that a membrane is permeable to a substance without testing our hypothesis.

CELLULAR AGGREGATION

A central goal of scientific work is to develop models that successfully predict the behavior of a system. We also ask that our models explain as wide a range of behavior as possible, ideally

5

MOTOR PROTEINS

Motor proteins are *mechanochemical enzymes* that convert energy from hydrolysis of nucleotides to *mechanical force*. This mechanical force is used to do work moving intracellular components (epg., *chromosomes*, *cytoskeletal elements*, *synaptic vesicles*, *mitochondria*), cells, tissues, and organisms. Generally, motor proteins bind a structure such as a mitochondrion and move it relative to a *cytoskeletal structure*-for example, a microtubule-which is tethered to other *cytoskeletal elements* or to the underlying cell substrate. Motor proteins play roles in events as diverse as chromosome segregation, nuclear migration and fusion, and membrane-bound organelle transport, flagellar beating, cell migration, muscle contraction, and *morphogenesis*.

Considerable *heterogeneity* exists in motor proteins. Despite differences at many levels from primary sequence to *enzymology* and *cell physiology*, common features allow motor proteins to be grouped into three families: kinesins, dyneins, and myosins. Variations both within and between families generate specificity needed for motor proteins to carry out a diverse set of functions, while similarities provide a common mechanism of action. Study of *mechanochemical enzymes* provides a basis for understanding cellular processes mediated by motor proteins, including intracellular maintenance and repair, *cell division*, *embryogenesis*, and *cell migration*.

CELLULAR AND SUBCELLULAR DYNAMICS

Intracellular components must be transported from one site to another because the internal milieu of *eukaryotic cells* is highly organised, with various cellular domains carrying out specialised functions. For example, DNA replication and transcription, protein synthesis, and energy metabolism are each restricted to specific cellular compartments. Such compartmentalisation is essential for biological function, but it imposes certain requirements. Cellular materials must be moved from sites of synthesis to sites of utilisation, with the result that the intracellular environment undergoes constant rearrangement.

For example, during mitosis chromosomes must replicate, line up at the metaphase plate, then migrate toward the two poles of the cell to form daughter nuclei. For daughter cells to receive equal genetic complements, such chromosome movements must occur in a precise and reproducible manner. Other specialised functions, such as contraction or locomotion, also require extensive intracellular reorganisation. Thus, muscle contraction entails an active rearrangement of the entire cytoskeleton. *Protists* and some cells in multicellular organisms (*embryonic stem cells*, *lymphocytes*, etc.) can migrate through a liquid medium or soft tissue. Such movements require continuous and substantial change in structures like *filopodia* and *cilia*.

The movement in all these examples requires generation of force. Motor proteins evolved to perform this role by *hydrolysing nucleotides* and converting the resultant energy to mechanical work. Considerable diversity exists in the types of work to be performed, and this functional diversity is reflected in the existence of motor proteins that differ in physical, enzymatic, and physiological properties. The goal here is to identify defining characteristics of motor proteins, and to show how variations in motor proteins allow them to carry out a wide array of functions.

FAMILIES OF MOTOR PROTEINS

Mechanochemical enzymes can be divided into three families: *kinesins*, *dyneins*, and *myosins*. Categorisations are based on common properties shared by all members of a family. These "*signature*" traits not only delineate the three motor protein families but serve as a framework to understand novel motor proteins.

Motors Proteins

Two criteria must be satisfied to qualify a *polypeptide* as a motor protein: it must be able to obtain energy via *nucleotide hydrolysis*, and it must use this energy to perform mechanical work. Since many enzymes can *hydrolyse nucleotides*, the ability to convert such energy to mechanical work is particularly critical. Historically, however, this ability has been difficult to demonstrate directly at the molecular level.

A variety of assays exist to demonstrate whether a polypeptide has *nucleotide phosphatase* activity. In vitro assays employing purified enzyme are the most *rigorous*. Typically, nucleoside triphosphate radiolabeled on the y-phosphate is added to an incubation mixture, and phosphatase activity is measured by release of radiolabeled *orthophosphate*. Different radiolabeled nucleotides can be added to determine the specificity of the enzyme's nucleotide phosphatase activity. All known motor proteins are ATPases. Basal ATPase activity is low in all three motor protein families but dramatically increases in the presence of an appropriate cytoskeletal element.

In vitro assays are also used to demonstrate force generation and performance of work by motor proteins. In the most common assay, the investigator attempts to pass cytoskeletal structures across a microscope coverslip coated with the motor protein to be tested. Depending on the motor, either microtubule or microfilament segments are added to the coverslip along with ATP. If the polypeptide has *mechanochemical* activity, the microtubules or microfilament will glide across the coverslip in the presence, but not in the absence of ATP. Movement of cytoskeletal elements can be observed using video-enhanced contrast/differential intereference contrast (VEC-DIC) or *video fluorescence microscopy*.

These in vitro assays are limited in that they document the potential of a protein to perform work but cannot identify the physiological role of a motor. To this end, alternate approaches must be which combine *in vivo* analyses with specific functional probes (*antibodies*, *pharmacological agents*, etc.) or genetics (*mutant phenotypes*, *antisense oligonucleotides*, etc.). Representatives of each family of motor protein have been found to satisfy both criteria, but only a fraction of all *putative motor proteins* identified to date have been characterised in sufficient detail to permit definition of their biological functions. Since few polypeptides have been rigorously proven to be motor proteins, the focus will be on selected, well-characterised examples from each motor protein family. Other less well-characterised members of each family exist, however, and new members are continually being discovered. The total number of distinct motor proteins in a given cell remains a matter of speculation.

Background and Discovery

Myosin, the first motor protein to be identified, initially was characterised as the mechanochemical enzyme of muscle fibers, where it is the main source of force generation during muscle contraction. The myosins possess a low basal ATPase activity, which is substantially enhanced by the presence of actin microfilaments. All myosins exert their force on actin filaments, not microtubules. Although myosins originally were described in striated muscle, other such proteins have been identified in both muscle and nonmuscle cells. At least seven distinct classes of myosins have been identified, which play a role in a diverse array of actin-based motility events ranging from muscle contraction to organelle motility.

The two best characterised classes of myosins are myosin I and myosin 11. Myosin II includes the myosins of smooth and striated muscle, as well as in nonmuscle cells, and dimerises to form a two-headed rod; these rods associate to form thick, bipolar filaments. Myosin II is thought to play similar roles-enhancing contractility and organizing the *cytoskeleton*—in all cell types. Myosin I differs from myosin II in both size and function. Myosin I polypeptides are about half the size of myosin II and do not form bipolar filaments. Some myosin I molecules may interact with membranes and play a role in the movement of membrane-bound organelles or the plasma membrane along actin filaments. The distribution and functions of the other five myosin classes are less well understood.

Dyneins were the next motor protein family to be identified. In the 1960s, dyneins were shown to be the motor protein responsible for axoneme or ciliary movements. *Axonemal* or *flagellar dynein* forms the outer and inner arms in 9+2 microtubule structure of cilia and flagella. In the presence of ATP, dynein causes flagella to undergo a whiplike motion, which may serve to propel ciliated cells (protists and sperm) through a liquid, clear debris from the laryngeal pharynx, and facilitate migration of specific stem cells during embryogenesis. Multiple dynein isoforms are found within a single flagellum.

Another dynein family member, MAP1 C dynein, was discovered in 1986. In contrast to axonemal dynein, which is found only in cilia and flagella, MAP IC dynein has a cytoplasmic localisation and is known as cytoplasmic dynein. *Cytoplasmic dyneins* may play a role in the movement of membrane-bound organelles along *microtubules* and possibly in the movement of *cytoplasmic microtubules*.

Kinesin was the last type of motor protein to be discovered. *Kinesins* were first described during studies of membrane-bound organelle transport in neurons. The axon is an extremely dynamic cellular domain, filled with organelles and structures in a state of continuous movement. Some organelles move in the anterograde direction (from the cell body toward the synaptic terminal) and other move in the retrograde direction (from terminal back to cell body) at rates up to 400 mm/day (2-4 p.m/s). Pharmacological manipulations demonstrated a requirement for ATP and microtubules in organelle movements.

Introduction of a *nonhydrolysable* analogue of ATP, adenylylimidodiphosphate (AMP-PNP) had a remarkable effect on the motility of organelles, freezing them in place on the microtubules. This observation not only demonstrated the existence of a new type of mechanochemical ATPase but provided a strategy to isolate the ATPase polypeptides. The ATPase bound to microtubules in the presence of AMP-PNP and was released from microtubules by ATP. In 1985 this strategy was used to purify a set of polypeptides that met all criteria for a *mechanochemical enzyme* involved in organelle movement in axons. This ATPase, kinesin, existed in a wide variety of *nonneuronal cells*, as well as in *neurons*. Soon thereafter, a number of proteins containing sequences homologous to the kinesin motor domain, but otherwise distinct, were identified. Discovery of these kinesin-related proteins generated the idea that the kinesin initially discovered in neurons defines a superfamily of related polypeptides. Members of the kinesin superfamily play roles in organelle transport, mitosis and meiosis, and possibly other microtubule-based motility events.

Diversity Among the Motors Proteins

Members of the three motor protein families are characterised by a distinct set of properties, shared by each other member of a given family. These "signature" properties can be used to distinguish between the three families and to define the class of motor molecule responsible for a particular form of motility. Variations in composition and sequence confer specificity and allow diversity in the functional roles of motor proteins. In the sections that follow, kinesins, dyneins, and myosins are described and compared in order.

Protein structure and composition

As originally described, kinesin is a heterotetramer of 380 kDa with two heavy chains (115-130 kDa each) and two light chains (62-70 kDa each). Electron microscopy shows that kinesin is an 80 nm long rod-shaped molecule, with three structural domains: globular heads, rod-shaped stalk, and fan-shaped tail. Two globular heads can be detected, each of which contains both ATP- and microtubule-binding sites. The rod-shaped stalk has a kink in the middle and separates the globular head domain from the fan-shaped tail. The tail domain is thought to be involved in binding to membrane surfaces (see Figure

1). Kinesin-related proteins are diverse in composition, molecular weight, and dimensions, but all appear to include one or more heavy chains with a globular motor domain.

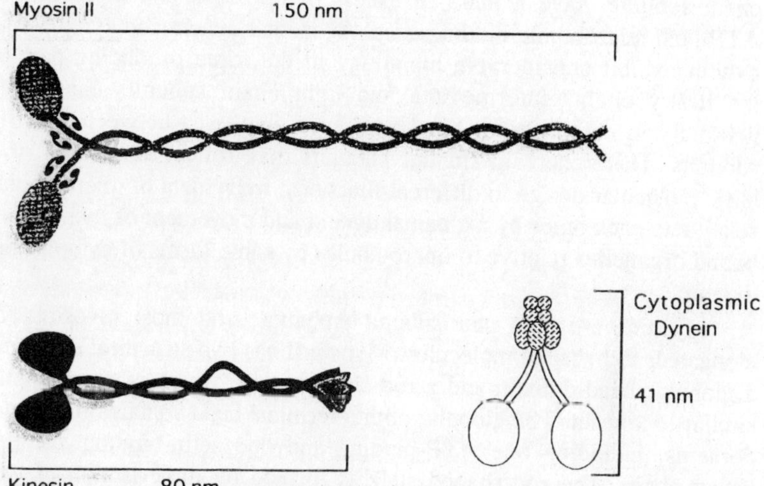

Figure 5.1 : Representative members of the myosin, kinesin, and dynein families of motor proteins. All three motor proteins have multiple suhunits. Myosin 11 is comprised of two heavy chain polypeptides each with a globular head and a rod region, which interacts with a second heavy chain to form an α-helical coiled-coil shaft.

Kinesin globular heads are formed by the heavy chain amino termini, while the heavy chain midregions dimerize to form an (Xhelical coiled-coil stalk. The fan-shaped tail is formed by the heavy chain carboxy-termini and two light chain subunits. Cells contain multiple isoforms of both heavy and light chains of kinesin, but the purpose of this *biochemical diversity* is uncertain. Isoforms of the light chain subunits of kinesin may target kinesin to different subclasses of membrane-bound organelles. Kinesin-related proteins may have their motor domains at the amino terminus like kinesin, at the carboxy-terminus, or even in the central region of the polypeptide chain. Little is known about the presence of light chains in kinesin-related proteins, but they are widely assumed to exist.

Like kinesin, dyneins are multisubunit enzymes, but they differ in structural organisation. *Axonemal dyneins* fall into two classes: two-headed enzymes (- 1200 kDa) and three-headed enzymes (-2000 kDa). Cytoplasmic dynein or MAPIC dynein is similar in size, *morphology*, and mass to two-headed *axonemal dynein*, but it differs in subunit composition. In both two- and three-headed enzymes, heavy chains form globular heads and a thin stalk, which connects to basal domains containing multiple intermediate/light chain subunits. Heavy chain subunits have a mass in excess of 500 kDa and contain both ATP- and microtubule-binding sites. Axonemal dynein and cytoplasmic dynein exhibit considerable homology in the force producing part of the heavy chain. Intermediate and light chain subunits can differ markedly in composition, but *homologies* may exist between specific subunits. Differences in smaller subunits may reflect adaptation of a basic molecular design to different functions: movement of microtubules relative to each other by axonemal dynein and movement of membrane-bound organelles relative to microtubules by some forms of cytoplasmic dynein.

Myosins are also multisubunit enzymes, and most myosins are elongated, rod-shaped molecules. Myosin II has two structural domains, a globular head domain and a rod-shaped stalk, which is superficially similar to kinesin. The globular amino-terminal head contains functional domains, including one ATP-binding and one actin-binding site per heavy chain. The rod-shaped stalk is formed by dimerisation of two heavy chains interacting in a coiledcoil a helix, like kinesin. Unlike kinesin, myosin II tails can interact to form large bipolar myosin thick filaments. The myosin 11 unit motor contains six polypeptide subunits: two heavy chains (-200 kDa) and two light chain pairs (20 and 16 kDa), with one of each pair of light chains per heavy chain head domain. In contrast, myosin I unit motors have only one heavy chain (- 100 kDa) and one light chain that is variable in size. *Homologies* to myosin II motors are restricted to the head motor domain containing both ATP- and actin-binding sites. Myosin I carboxy-terminal ends are specialised and may contain a nonnucleotide-sensitive actin-binding site, which would allow myosin I to cross-link actin filaments or a membrane-binding site. Myosin light chains are thought to be involved in regulating the mechanochemical activity of myosins. A variety of other myosins have been identified but are less well characterised. These proteins may exist as monomers, dimers, or multimers, but their biology and function remain to be defined.

Biochemical and biophysical properties

All three motor protein families have an ATPase activity, which is activated by the presence of the appropriate cytoskeletal partner, microtubules (kinesin and dynein) or microfilaments (myosin). In effect, motor proteins utilize ATP only in the presence of appropriate cytoskeletal elements and ATP as the physiologically relevant *nucleotide*. Maximum rates of *nucleotide* hydrolysis and translocation appear to be similar for the three families of motor proteins, but specific members may be much less efficient in in vitro assays. Motor proteins have characteristic pharmacological and biochemical properties, which may be used for identification. For example, the three families are similar but not identical in their ability to utilise nucleotides other than ATP.

A major difference between these motor protein families is the effect of AMP-PNP, the *nonhydrolysable* analogue of ATP. AMPPNP stabilizes binding of kinesin to microtubules but weakens binding of dynein to microtubules and myosin to microfilaments, indicating a difference in the enzymatic cycle of three families of motor proteins. In the case of kinesin, hydrolysis of ATP appears to be necessary for the release of kinesin from microtubules. In contrast, binding of ATP to dynein and myosin is sufficient to release them from microtubules or microfilaments, and hydrolysis occurs after release.

Another difference between the three groups is the effect of orthovanadate on ATPase activity. All three ATPases are inhibited by orthovanadate, but effective concentrations vary by a factor of 2 to 10. A more dramatic difference is that dyneins exposed to UV radiation in the presence of orthovanadate and ADP are sitespecifically cleaved, a property that has been used to implicate dyneins in cell activities. Myosins are subject to similar cleavage under somewhat harsher conditions, but kinesin does not appear to be subject to UV/ vanadate cleavage.

All three types of motor protein mediate directional movements along a *cytoskeletal* element. *Microtubules* and *microfilaments* are polymers formed from tubulin and actin subunits, respectively. Both polymers are polar structures (rapidly growing plus end and slowly growing minus end), and the orientation of each polymer is precisely regulated in cells. *Microtubule* and *microfilament* polarity may be determined in a variety of in vitro or in vivo situations, so the direction of translocations can be defined with respect to the polymer polarity. Based on in vitro gliding assays, each motor protein exhibits a characteristic directionality of movement. Many members of the kinesin and myosin families are plus-end-directed motors because they move toward the plus ends of microtubules and actin filaments, respectively. However, some kinesin related proteins are minus-end-directed motors. All dyneins examined to date move toward the minus ends of microtubules, and thus are minus-end-directed motors. Directionality of a specific motor protein is critical for understanding the role it may play in the cell.

CELLULAR ROLES PLAYED BY MOTOR PROTEINS

How many motor proteins does a cell need for survival and what roles do they serve? This question is particularly intriguing in light of the ever-increasing list of motor proteins in cells. A single cell, such as an intestinal epithelial cell, contains multiple members of all three motor protein families. Although cell-specific motors exist, each cell requires a diverse set of motors for normal function. Individual motor molecules may perform a single cellular function, or the functional roles of these proteins may overlap, with the result that one motor may be able to compensate for loss of another motor if necessary. Two characteristics determine the cellular function of a specific motor protein: the specificity of binding to cargo structures (*cytoskeleton*, *membrane organelles*, etc.) and the directionality of movement along a *cytoskeletal structure*.

The most variable regions of all three motor families are found in the "*tail*" domains, usually at the *carboxyl termini*. These domains are thought to specify what a given motor protein can move. Little is known about the nature of this interaction, but evidence exists for specific binding of specific motor molecules to *microtubules*, *microfilaments*, various membrane surfaces, and kinetochores. Associated light chains may convey the specificity, but motor heavy chains may also contain relevant targeting information.

The cytoskeleton of cells is highly organised. Microtubules originate near the nucleus of an idealised intestinal epithelial cell and radiate toward the cell periphery. All these microtubules are oriented with minus ends anchored at an organising center associated with the centrosome, so the plus ends are toward the periphery. In contrast, microfilaments are concentrated in peripheral domains (epg., in microvilli or on the plasma membrane). Membrane microfilaments are anchored at the membrane surface by the plus end, with minus end toward the interior. The situation in real cells may be more complex than shown in Figure elsewhere in this chapter, but the most common situations are represented.

Organisation and polarity of cytoskeletal structures influence the roles played by motor proteins. For example, the kinesin superfamily contains both plus- and minus-end-directed microtubule motors. Kinesin attached to membrane organelles moves them from the Golgi

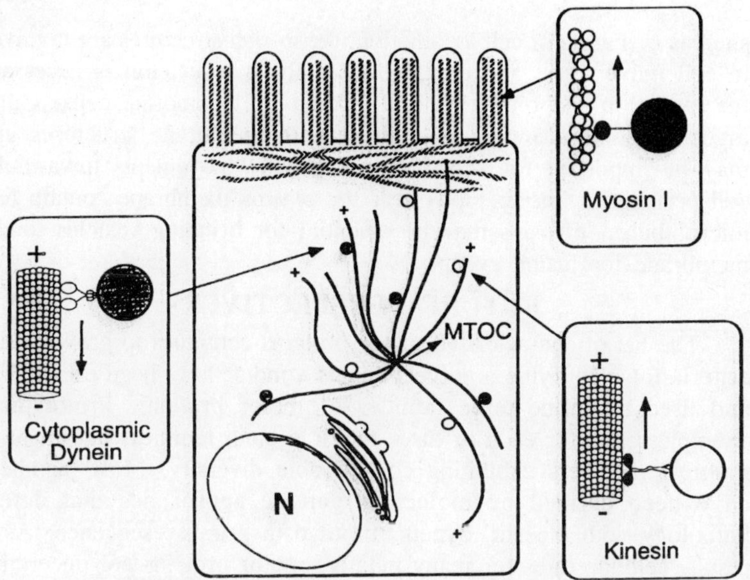

Figure 5.2 : Selected cellular roles for the three motor protein families in a specific cell type. Each intestinal epithelial cell contains multiple members of all three motor protein families, which serve a variety of roles in both dividing and interphase cells. Many functions benefit from the inherent polarity of cytoskeletal elements within the cell, which provides a basis for directionality of movements.

complex near the cell center to the periphery. This feature is important in cells whose membrane *organelles* must be transported considerable distances-in neurons, for example, organelles are transported down axons that may be a meter or more in length. Kinesin-related proteins with other binding characteristics play different cellular roles, such as mediáting chromosome migration from metaphase plate to spindle-pole bodies during mitosis.

Dyneins, like kinesins, are microtubule-based motors, but most dyneins are minus-end-directed motors. Axonemal dyneins produce ciliary beating by attaching stably to one microtubule and exerting a force on an adjacent microtubule in a flagellum. Some cytoplasmic dyneins interact with membrane structures, while others may attach stably to *microtubules*. *Cytoplasmic dyneins* appear to move membrane organelles like endosomes and lysosomes from the periphery to the cell center for recycling as well as potentially playing a role in mitosis and meiosis.

Myosins are plus-end-directed, microfilament-based motors. Myosins I and II are thought to be involved in contractile events, such as in muscle or cell locomotion. Actin-rich structures are involved in cell movements, and contraction of the cytoskeleton is necessary for extension and retraction of filopodia by locomoting cells. Other myosin I motors appear to interact with membrane structures and may be important for short-range organelle movements toward the cell periphery. Since regions near the plasma membrane contain few microtubules, myosins may be essential for bringing vesicles to the membrane for fusion events.

FUTURE PERSPECTIVES

The list of characterised motor proteins continues to grow. Clear criteria for identifying a polypeptide as a motor have been established and used to define three families of motor proteins. Prototypical examples exist for each of three motor protein families, but all have multiple members exhibiting considerable diversity. Most members have been defined by molecular genetic approaches that detect homology with kinesin, dynein, or myosin primary sequence. As a result, cellular roles for many putative motor proteins are uncertain, but a variety of novel roles have been proposed. The near future should produce characterisation of additional motors including physiological function, mechanisms of regulation, and the role of motor proteins in disease.

6

EXOCYTOSIS

There are thought to be natural processes by which invaginations in the cell membrane carry molecules into cells within vesicles formed from the invaginating membrane and molecules are also thought to be moved out of cells, or secreted, in an analogous fashion, usually envisioned today in the form of the exocytosis model. Both of these concepts are examples of a wide range of processes, hypotheses, and constructions that we call "*vesicular transport*," and that have in common the following notions:

(1) membrane-bound vesicles are either formed as the result of transport, as in invagination processes, or preparatory to its occurrence, as in the formation of secretion granules prior to *exocytosis*;

(2) transport is not driven by the kinetic energy of the transported molecules, but by an external force;

(3) transported molecules do not cross the membrane by passing through its substance or through *channels* in the membrane;

(4) transported molecules do not move individually, but en masse or in *bulk*;

(5) the membrane barrier must be removed for transport to occur;

(6) this involves the macroscopic separation of adjacent portions of membrane from each other; and

(7) this schism in the membrane occurs either during the insertion of additional membrane, or the deletion of membrane segments.

EXISTENTIAL HYPOTHESIS

What kinds of evidence might be of value in helping us evaluate such an hypothesis? In the most general sense, we can state that all specific or particular vesicular hypotheses, as well as vesicular hypotheses in general, are strictly existential, as opposed to universal, hypotheses. An existential hypothesis is one that cannot be proven false in any general sense by any single case for which we reject the hypothesis. For example, if we are unable to demonstrate the presence of a particular vesicular mechanism in one cell or another, who can say that it will not be found elsewhere? In the case of a universal hypothesis, a single negative case requires rejection of the hypothesis for all possible cases. For example, we only need one exception to the law of conservation of energy and mass to require its rejection. Since existential hypotheses cannot be rejected in general from a particular false experience, likewise they cannot be proven true in general from a specific example or case.

With this idea in mind. One can make the following statements about vesicular models:

1. Proof for the existence of one type of vesicular mechanism does not bear upon the existence of any other type of vesicular process. This is a matter for independent proof. The agreement that there is the possibility that a process might exist, based on the existence of another similar or analogous process, has no bearing on its actual existence. It only means that we should not reject the idea out of hand.
2. Even if a particular vesicular process can be shown to account for the transport of a molecule or group of molecules in one cell type, nevertheless, the existence of the same process in another cell in pursuit of the transport of different substances must be independently proven.
3. Proof that a process occurs under a particular set of experimental circumstances does not necessarily mean that it occurs as a natural process. "*Real*" processes can be produced by the investigator's experimental system or design, and therefore the mere existence of an event does not indicate that it occurs naturally or accounts for transport under normal circumstances. There must be specific evidence to support this contention.
4. Even if we can show that a particular vesicular process occurs naturally and transports a molecule of interest, we

cannot presume that it exists to perform the physiological function that we envision for it without evidence that our hypothesis is true in this regard as well.

5. Even if we can demonstrate that a particular vesicular mechanism occurs naturally and transports the particular substance of interest, we cannot assume that it accounts for its movement quantitatively without specific evidence concerning the quantitative aspects of its transport.

FALSIFICATION OF HYPOTHESES

Although the existence of some vesicular processes seems clear, for example, the phagocytosis of food by protozoans, evidence for many is quite fragile and is at times no more than an assertion by an investigator that it occurs. The evidence for most such hypotheses is *not* the *direct* observation of its occurrence in real space-time, as in *phagocytosis* where one can observe a particle being engulfed by a living cell in the light microscope. Usually it is indirect; observations that are consistent with such a mechanism, but that do not rule out alternative explanations for what is seen, nor that demonstrate its existence directly by measurement or observation of the natural continuum. The importance of this latter point will become clearer as we consider the evidence for exocytosis below.

For the moment, let us consider a point of scientific methodology that concerns the difference between "*seeking* evidence in support of a hypothesis," and "*testing a hypothesis*." Frequently, investigators "*seek support*" for a particular hypothesis or point of view, but do not actually "*test*" it. That is, one possible outcome of an experiment is not that the hypothesis will be rejected. At best "*support*" will not be forthcoming. We say that if our hypothesis is correct, then we would expect to observe this or that. However, the statement that the absence of the particular occurrence provides negative evidence for our hypothesis cannot be made. When this approach is used, all hypotheses are in a sense correct to one degree or another, and differ only in their relative degree of corroboration; that is, how much support exists for the particular hypothesis. Its weakness can perhaps best be appreciated by considering the real possibility that two totally antithetical hypotheses have been proposed to explain a particular phenomenon and that they are equally supported by different corroborative observations. Applying this reasoning, we would conclude that both hypotheses (a hypothesis and its logical antithesis) are simultaneously true, or that "*truth*" itself fluctuates between one reality and another depending upon a continuously changing balance of evidence. Thus, in order to determine satisfactorily to what degree a hypothesis reflects the real world, we should do more than seek support for it. We should attempt to test it, against other hypotheses, even antithetical ones, as well as in its own terms, as either being true or false. It is more powerful to be able to say that our hypothesis has been substantiated or corroborated by numerous tests of its validity, *its falsification being rejected,* than merely asserting that there is accumulated evidence that supports or is consistent with our view. We must be willing and able to prove our hypothesis false in order to prove it true: a more difficult task than it might seem at first glance.

We know intuitively that some experimental tests of hypotheses are more powerful than others. Such differences derive from the ability of a particular experiment to really test our hypothesis; that is, from the degree to which it is capable of proving our hypothesis false. Thus, an experiment that has a high falsifiability quotient, if you will, is a stronger test and hence a better experiment than one that has little potential for proving our hypothesis false. In addition to tests, hypotheses themselves can be more or less powerful depending in part upon how amenable they are to being proven false.

Some hypotheses are so vague or otherwise flawed that they are virtually impossible to prove false. Of course, such a hypothesis cannot likewise be proven true, even though one might amass considerable evidence in its support. Thus, to the degree that falsifying tests are not carried out, or are not possible or practicable, we cannot show that our hypothesis satisfactorily accounts for the observed phenomenon regardless of how much evidence we might marshall in its behalf. Only when a hypothesis is adequately falsifiable, and when we have performed numerous tests that could have, but did not, lead to its falsification, do we have the right to accept it as a satisfactory approximation for the moment.

This issue of the "*strength*" of tests and hypotheses should be borne in mind as we move deeper into particular experimental issues and ideas. In addition to the issue of falsification, we should also keep in mind the existential nature of the exocytosis hypothesis, that is, the need to prove its existence system by system, and cell by cell. We cannot generalize from one cell type to another, nor accumulate evidence for the process in general by adding that obtained in one system to that obtained in another.

EVIDENCE OF FORM

Although evidence for exocytosis varies in regard to both type and degree of support from cell type to cell type, the preponderance is of one particular kind that we can call "*evidence of form*"; that is, observations of particular static shapes in cells that are viewed as being consistent with the occurrence of exocytosis in real space-time.

In recent years there have been some attempts to prove the existence of such processes in cell-free systems by demonstrating that fusion of granule and plasma membrane can occur under certain conditions. Such experiments may demonstrate that fusion can occur between two membranes, but that the relationship between such an observation and the existence of exocytosis in an intact cell is tenuous. Perhaps we can agree that it is possible under a variety of conditions for biological membranes to fuse with each other. The issue is not this capability, although there are questions concerning the potential for its occurrence in certain specific situations *in situ,* but its relationship to the existence of a natural process called exocytosis. Perhaps someday conclusive evidence for *exocytosis* will come from such a purely "*reductionistic*" approach, whereby one attempts to prove the existence of exocytosis by elucidating its underlying detailed mechanism.

SECRETORY GRANULES IN CELLS

Secretory granules are found in many, but not all, cells that are known to secrete proteins and other organic compounds such as *neurotransmitters*. It has been established that many of these granules contain the product or products that are secreted, although it has not been established that all of their contents are secreted in any case to my knowledge.

The mere existence of granules in cells can be taken as evidence of exocytosis. The reasoning is that "granules are involved in exocytosis where it is thought to occur, and therefore their presence, particularly if they contain a product that is secreted, is prima facia evidence for exocytosis in a cell." Of course, the presence of granules suggests the possibility that exocytosis might occur, but it provides no evidence, of even a weak variety, that it does, even in the most granule-laden cell.

If we were to argue that the presence of granules in cells is prima facia evidence for an exocytosis process, then wouldn't their absence be evidence that the hypothesis is false? Many cells that are known to secrete large quantities of proteins, such as the parenchymal cells of the liver, which contain few if any true *secretory granules*, are still proposed to carry out transport by an exocytosis mechanism similar to, if not identical with, that proposed for cells with granules. Their absence is merely explained by ad hoc hypotheses; either there is little storage of product within the cell and hence there are very few granules, or the rate of granule turnover is so rapid that few granules are required to account for the observed secretory rate, or both, and so forth.

Of course, the presence of granules in cells is evidence for *exocytosis* in the same way that the presence of the sun in the sky or a rock on the ground is evidence that it moves. This is not to say that granules are not involved in such processes in some cells or under some circumstances, nor that their presence in cells does not suggest the possibility or hypothesis that such a process might exist there, but their presence is simply not evidence for one type of transport process or another. The presence of granules in cells is evidence for and proof of the presence of granules in cells, and nothing more.

DISAPPEARANCE OF GRANULES

The number of granules in micrographs of secretory cells may decrease when the secretion of a product contained within them is augmented, although such a decrease is not always seen. This disappearance, is often taken as evidence for *exocytosis*. "How else could they disappear?" it would be asked.

The disappearance of granules during augmented secretion was first shown for the *acinar cell* of the pancreas by Rudolf Heidenhain in the mid19th century. Heidenhain found that under certain conditions of augmented secretion the number of granules in the cell, as viewed through the light microscope, appeared to be diminished. Of course, this observation in itself did not necessarily mean that granules left the cell in toto. Heidenhain was aware of this because both he and his contemporaries, Kiihne and Lea, had also observed that the size of granules in the tissue decreased simultaneously. These size changes suggested an alternative to the release of granules in toto; that is, a decrease in size to their actual, or merely apparent, disappearance within the cell.

Apparent disappearance, because if granule size is decreased, even if their number remains unchanged, micrographs may well display fewer granules. For example, the *zymogen granules* of the pancreas average a little less than 1 μm in diameter in the fasted state, which is close to the theoretical resolution of *light microscopy* (the wavelength of visible light being -0.5 μm), and if granule size decreases by half when secretion is augmented, then the resolution of the light microscope starts to become limiting; and small, although still present granules, may not be seen in the microscope and may appear to have disappeared. The same effect occurs at the level of electron microscopy, although for different reasons. In this case it is not that resolution is limiting, but that the probability of sampling a particular granule in any given tissue section decreases as size decreases [number/unit volume= number/unit area × (mean diameter)$^{-1}$]. Ermak and I have found that large apparent changes in granule number may "*dlsappeur*" when thc probability of sampling granules of various sizes is taken into account

Therefore, the apparent disappearance of granules can be explained in ways that include but are not limited to their loss from the cell in *toto*. And for this reason, we cannot argue that such "*loss*" or "*disappearance*" is evidence in favor of a particular mechanism of secretion, any more than we can argue that the presence of granules in cells is evidencc for such a process. As with their presence, the disappearance of granules can only be proof of their disappearance.

Table 6.1 : The Effect of Granule Size on Estimates of Granule Number

Time Period (Days Postcoitum)	*Mean Granule Diameter (μm)*	*Number of Granules/Volume (Arbitrary Units)*	
		Uncorrected	*Corrected for Size*
17	0.55	1	1
20	1.22	3	1.36

We can consider the negative case here as well. If secretion can be greatly augmented without a decrease in real granule number, does this observation falsify the exocytosis hypothesis? That is, do we conclude that since secretion occurs in the absence of a change in granule number, that the *exocytosis hypothesis* is false? Proponents of the hypothesis would be quick to point out that constancy of number might merely indicate that the rate of formation of new granules is able to keep pace with their rate of loss from the cell. A decrease in granule number is proof for exocytosis in the same way that the absence of a change in number is proof that it does not occur.

EXOCYTOSIS

The most frequently cited evidence for the existence of *exocytosis* is the exocytosis or *omega figure*. It is an outline of the surface membrane of the cell that has a form that is consistent with the prior occurrence of exocytosis; that is, it could be a geometric sequelae to exocytotic fusion. These forms are called because they are shaped like the Greek letter omega (f1), the circular portion being thought to represent the membrane of the secretion granule fused to the cell membrane and the arms of the figure the cell membrane proper. Omega figures are seen from time to time, more frequently in some secretory cells than others, and are sometimes very evocative of the idea that an exocytosis-like event has occurred, although in many cases they are not very convincing. However, as with the presence of secretion granules in cells or their disappearance from them, regardless of the visual impact that a particular form may have, the mere presence of the omega figure does not indicate that exocytosis has taken place. Although it is true that if exocytosis had occurred such a *geometric* figure might well be seen in the outline of the cell for a period of time, and hence its existence is consistent with such a process, the prior occurrence of exocytosis is only one possible explanation for its presence. Often the outlines of omega figures cannot be or are not distinguished from chance or other geometric variations in the outline of the cell surface; or they may be real enough, but the result of an artifactual, rather than a natural, fusion of membranes, for example, produced by the fixation of tissue or other steps in the preparation of tissue samples for viewing. Thus, it is an hypothesis that the omega figure is related to the prior occurrence of a natural process of exocytosis. As with the other evidence of form, omega figures can only be proof for the existence of omega figures.

The weakness of the omega figure as evidence for a *dynamic process* is compounded by the fact that they are usually sought by investigators; that is, they tend to be infrequent occurrences that are discovered as the result of a search. If they are found, then the exocytosis hypothesis is thought to be supported by the observation, but if they are not, the search continues by increasing the size of the sample surveyed or by looking for better circumstances and conditions. Thus, the negative observation (the absence of omega figures) does not lead to the rejection of the hypothesis, but to a continued and perhaps expanded search to validate it. In the end we can only view the *omega figure* as significant evidence for exocytosis if we *assume* that exocytosis is the reason for its presence, and of course to assume that is to assume the existence of exocytosis in the absence of evidence.

If we wish to show that the omega figure is more than a chance, artifactual, or other occurrence either unrelated or only indirectly related to natural secretion, then we must provide a substantive connection between such figures and the secretion of product. First and foremost, if the omega figure is related to secretion, then its frequency should be related to secretory rate. If the secretory rate increases so should the frequency of omega figures. Indeed, the frequency of omega figures would be expected to increase proportionately. Although this correlation would certainly help place the observation of the omega figure on a firmer scientific footing, we would still have to demonstrate in addition that the observed frequency of such an event accounts quantitatively for the amount of material that is secreted; that is, the fusion of so many granules leads to the release of so much product. Unfortunately, even if we were able to show this, we still could not, despite the correlation and obvious intuitive connection, say that the omega figure proves the existence of exocytosis because we still would lack direct evidence that such a process occurs in real space-time and causes these forms.

Perhaps we can appreciate the weakness of such evidence better by considering the negative case once again. When an increase in the frequency of the omega figure is *not* observed when the rate of secretion is increased, as seems to be the case for the acinar cell of the pancreas and many other cells in which exocytosis is thought to occur, it is sometimes argued that exocytosis is too rapid for us to "*catch.*" But of course, if there are "*real*" omega figures in the unstimulated state, then we should see proportionate changes with increased rates of secretion however fast the process may be. Thus, if the omega figure is in fact a sequel to secretion by exocytosis, then shouldn't an increased rate of secretion necessarily lead to an increase in their observed frequency? If this is not seen, then would we not be required to reject the hypothesis? There are numerous instances where on this basis, rejection would appear to be required by the available evidence. However, it is not generally forthcoming.

The reason for this is that even in the absence of such a correlation, we cannot necessarily reject the exocytosis hypothesis, because exocytosis might either occur without a discernable *omega figure*, or the duration of its appearance might be reduced when the rate of secretion is increased and so forth. Thus, these figures, even when they increase in frequency during augmented secretion, are proof for exocytosis in the same way that their absence, or the absence of changes in their frequency, is proof that such an event does not occur.

MEMBRANE SURFACE AREA

Another *deductive hypothesis* derived from the *exocytosis* construct proposes that the surface area of the plasma membrane should increase if the rate of exocytosis is increased, because additional membrane is being inserted into the plasma membrane of the cell. Such increases have been observed, although this is not by any means a universal occurrence in cells thought to secrete in this fashion. At first glance, this type of evidence may seem to make a "*dynamical*" connection that much of the evidence discussed above lacks. However, as with the other evidence of form, an increase in surface area of course does not necessarily indicate the prior occurrence of exocytosis. There may be other explanations, most obviously, an increase in the volume, and hence the surface area, of an actively secreting cell. The case for a dynamic connection is more convincing if large disproportionate increases in surface area occur, particularly if localised to a particular cell surface, as appears to occur when HCl secretion from the oxyntic cell of the stomach is augmented with stimulants. Thus, if we can show a change in surface area due to alterations in the rate of secretion we still need to be able to ascribe it to our proposed mechanism, as opposed to other causes, either unrelated, indirectly related, or related to secretion in another way. Also, as with the other examples that we have discussed, when increases in surface area are not observed, the hypothesis need not be rejected. Many cells that are thought to secrete by exocytosis do not show this effect, and our hypothesis can be saved by the ad hoc hypothesis that the rate of membrane reuptake by the cell is equal to the rate of its insertion, and hence a steady-state pertains. Of course, this kind of reasoning allows us to argue that a decrease in the surface area of a cell coincident with an increased secretory rate is also evidence for exocytosis; an increase in the rate of exocytosis leading to an even greater increase in the rate of membrane uptake into the cell.

PARTICLES

Patterns of particles, which are probably membrane-associated proteins, are sometimes observed in freeze-fracture images of membranes taken from cells thought to secrete by exocytosis and have been proposed to be evidence for its occurrence. The *particles* (proteins) would be directly involved in or otherwise related to fusion between granule and cell membrane. This proposal is analogous to the other evidence of form that we have discussed. In itself it is merely an hypothesis for the function of the particles and is not evidence for exocytosis. This confusion between hypothesis and evidence is further compounded when investigators themselves order independently obtained micrographs that contain different patterns of particles in an attempt to convey the impression of a sequence of events occurring in real space-time, as in the formation and subsequent dissolution of rosette or linear patterns of particles. Of course, this is merely a visual display of the investigator's hypothesis since he or she has no idea of how the different forms are really related to each other, if indeed they are related at all; no less if they are related to secretion and in the way proposed.

If fusion of granule and cell membrane takes place and produces a hole in both membranes, as the *exocytosis hypothesis* proposes, then these holes should appear as pits in the membrane or in half membranes viewed in *freeze-fracture* replicas. This observation is quite similar to the omega figure and the same types of issues apply. Their presence may be due to events other than the proposed secretory mechanism, and their absence can be explained by ad hoc assumptions concerning the rate of the secretory process. And their discovery, like that of the omega figure, is usually the result of a search for an infrequent form.

GENERAL DISCUSSION

The weakness of evidence of form and the ambiguity and confusion that it produces between what is evidence and what is hypothesis too often goes unappreciated. The examples that has been given are sufficient to show both the kind of "evidence" that has been mustered in support of the exocytosis hypothesis and the deep problems with it.

One cannot emphasize enough the need to correlate such cytologic observations with the presumed functions in a rigorous and quantitative manner. However, the central weakness of essentially all of the evidence that has been brought to bear on the exocytosis hypothesis rests inescapably on the fact that the secretory process itself is never measured or followed in real space-time independent of the hypothetical construct of exocytosis. It is for this reason that it is so easy to propose ad hoc hypotheses to account for negative results, and it is also for this reason that evidence and hypothesis are so easily confused with each other. Perhaps this can be highlighted by considering an experiment that attempts to study the release of acetylcholine at the cholinergic synapse. It is widely, although not universally, believed that the exocytosis of acetylcholine-containing vesicles occurs at the synapse and accounts for the transmission of the impulse. However, this process, as all other proposed "*exocytoses*," has never actually been observed in living cells. In the past, even evidence of form in support of the exocytosis hypothesis at the cholinergic synapse was rather weak. Synaptic *transmission* occurs very rapidly and perhaps, it was thought, the exocytosis event was too rapid to be seen by the usual approaches and special measures would have to be taken to "*see the process*" as it occurred. In the particular experiment at hand, the investigators proposed to stop exocytosis in midstream by rapidly freezing the tissue and then looking for evidence of its occurrence, most importantly, pits in freeze-etched membranes. They were able to find them under certain circumstances (that is, in the presence of an exogenous chemical that "*helps*" increase their apparent frequency) and from these observations concluded that they had provided direct, even kinetic, evidence for exocytosis in this system. But what kind of evidence was it? What was stopped in midstream? It was not exocytosis at all, but rather the "*hypothesis*" of exocytosis. That is, a known process was not stopped; indeed, the investigators sought proof of its existence. Thus, a hole in a membrane cannot be proof that exocytosis occurs as a natural process, although it is certainly consistent with it. The hole may be due to other unrelated or artifactual causes that we have no way of knowing.

As discussed earlier, the problem is simply that we cannot ask whether or not the secretion process *is* behaving as our mechanistic hypothesis predicts, either kinetically or thermodynamically, because we are not actually following that process. Even if we confirm a particular prediction of our hypothesis, we cannot say with certainty that the observation is related to the transport process that we think we are investigating, because its behavior has only been characterised in terms of the hypothesis itself. The question that we must ask then is what, if anything, our observations tell about the secretory process independent of the truth or falsehood of our hypothesis. That is do our observations have an existence as descriptions of the events of secretion independent of our hypothesis? For evidence of form in general, the answer is negative and observations are only descriptive of secretory events if the hypothesis is correct.

EXOCYTOSIS IN THE PACNCREAS

However weak this evidence may seem, it should be noted that it is the unusual case (if one exists) in which all types of evidence of form can be shown for a single proposed exocytic process. For many years the pancreas was cited as the system in which the evidence for exocytosis was most prepossessing, and indeed it has been used to argue for the existence of such processes in other cells where the evidence was not thought to be as secure.

20 years ago the idea that exocytosis accounted for the secretion of the digestive enzymes by the pancreas was already well established, and textbooks and monographs assured us, as they still do, that this was the way in which things happened. The process was called "*reverse pinocytosis*" at that time; but no matter, the proposed mechanism was essentially the same. However, a search of the primary literature for evidence left one feeling that something was missing. The evidence appeared to be threefold:

(1) the cell contained secretion granules, zymogen granules, which were known to contain the enzymes that were secreted;

(2) these granules seemed to disappear from the cell during augmented secretion under certain circumstances; and

(3) the *omega figure* had been reported.

The following is a brief dialogue that occurred between Prof. Neaverson Hokin had presented a paper on the subject of protein transport by the pancreas at a symposium on exocrine glands in 1968. I think that it sums up the situation quite well. Prof. Neaverson-Hokin asked, "If (I had) really examined evidence that reverse pinocytosis occurs at all in the pancreas?" I responded weakly, "The evidence for reverse pinocytosis is based on two sorts of things (being terribly precise). One is electron micrographic pictures (of omega figures) which are ..." She interrupted me, "Is it pictures or picture?" I responded, "Pictures-there are several." But her point was made. There may have been more than one picture, but not much more, and the published evidence was certainly minimal. Proponents of the exocytosis theory argued that such figures were seen frequently, but this was merely hearsay, and the frequency of their occurrence was really unknown and apparently unstudied.

In the years since this exchange, the cisternal packaging–exocytosis paradigm has become even more generally accepted as *the* mechanism for protein secretion by the pancreas, and exocytosis is believed by most scientists in this and related areas to account for the final step in the transport process. Yet the evidence for its existence has remained essentially unchanged. Several new omega figures have been published since 1968, but little else has been forthcoming. There does not appear to be an increase in the occurrence of such forms in actively secreting cells. There does not appear to be a change in the surface area of the cell membrane or its apical surface. Freeze-fracture evidence of pits or particles has yet to be published to my knowledge. The fact remains that evidence for the existence of exocytosis in the pancreas, no less evidence that such a mechanism accounts quantitatively for the natural process of secretion during the digestion of a meal, is weak at best.

7

ENZYME MECHANICS

Transient state kinetic analysis involves following the time course of a reaction to completion for two reasons: to determine the rate of the reaction and to use that information to establish mechanism. The term *transient state* usually refers to the rapid (millisecond time scale) analysis of chemical or enzymatic reactions in the early phase of the reaction preceding the more commonly studied *steady state*. However, most of the kinetic analysis of biological reactions relies on transient state kinetic methods because these methods pertain to single molecular events. For example, the binding of a ligand to a cell surface receptor and the binding of a repressor to DNA fall under the realm of transient kinetics because both involve a single reaction proceeding to completion, as opposed to the multiple turnovers that characterize the conceptual framework of steady state enzyme kinetic analysis.

The contribution describes the fundamentals of transient state kinetic analysis of bimolecular and unimolecular reactions in biological systems.

THEORETICAL ASPECTS

A reaction involving the binding of a substrate S to an enzyme E is described by the kinetic equation:

$$E + S \underset{k_{-1}}{\overset{k_1}{\rightleftharpoons}} ES$$

The constant k, defines the *second-order rate constant* for substrate binding to the enzyme site, while the value of k_{-1}, defines the *firstorder rate constant* for dissociation of substrate from the enzyme site. A first-order rate constant gives the rate of spontaneous reaction in units of events per second (s^{-1}). A second-order rate

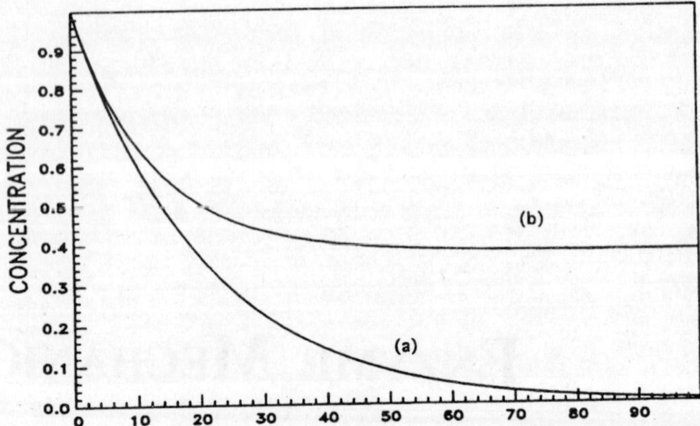

Figure 7.1 : The time courses two hypothetical enzyme-substrate binding reactions: (a) irreversible and (b) reversible.

constant includes the concentration dependence in units of events per second divided by concentration ($M^{-1}\ s^{-1}$). For most biological reactions, second-order binding rates fall in the range of 10^4-10^9 $M^{-1}\ s^{-1}$, where the upper limit is a function of the diffusional collision frequency. Dissociation rates are typically on the order of 10^{-1} - 10^3 s^{-1} (half-lives of 2 to < 1 m).

If the binding of E to S is irreversible ($k_{-1} = 0$), and the concentration of S is greater than the concentration of E, the reaction will proceed to completion at a rate of $k_{obs} = k_1[S]$. The time course of disappearance of E will follow *a single exponential* equation, $[E] = [E]_0 \exp(-k_1[S]t)$. Analysis of the reaction time course by nonlinear regression would give the value of k_{obs} (s^{-1}). Division of the observed rate by the concentration of substrate gives the value for the second-order rate constant ($M^{-1}\ S^{-1}$).

In practice, reactions in biological systems are rarely irreversible, so we must consider the kinetics of the reversible reaction, where the magnitude of k_{-1}, is comparable to k_1 [S]. Under these conditions, the time course still follows a single exponential, but the observed rate constant k_{obs} is $k_1[S] + k_1$. Because the data analysis involves fitting to a single exponential, the observed rate of approach to equilibrium is equal to the sum of the forward and back rates. The time course of reaction is given by the equation:

$$[E]-[E]_\infty = ([E]_0 - [E]_\infty)\exp(-k_{obs}t)$$

While the rate of the observed reaction is increased by the rate of the reverse reaction, the amplitude of the observed reaction is reduced by the extent of reaction at equilibrium, where $[E]_\infty$ is the concentration of free enzyme at infinity.

This elementary kinetic analysis provides the fundamental conceptual basis for examination of all biological reactions. Additional complexities arise as a result of the addition of multiple steps along a reaction sequence, and each step in a pathway contributes a single exponential to the equation describing the process. For example, a conformational change may occur in the ES complex leading to tighter binding (*E'S* complex):

$$E + S \underset{k_{-1}}{\overset{k_1}{\rightleftharpoons}} ES \underset{k_{-2}}{\overset{k_2}{\rightleftharpoons}} E'S$$

The time dependence of the two-step binding reaction follows an equation involving the sum of two exponential terms. In the most simple case involving two irreversible reactions, the rates of the two phases of the reaction equal the rates of the two steps (k_1 and k_2). However, in the general equation allowing for reversible reaction steps, the magnitude of each of the two observed rates will be a function of all four rate constants. Different results can be obtained, depending on the signal that is being measured and the relative rates of each reaction.

Data analysis can be accomplished by fitting data to integrated rate equations by nonlinear regression. However, the solution of rate equations for complicated pathways is tedious and often requires the use of simplifying assumptions. The use of computer simulation by numerical integration of the rate equations overcomes the limitations of conventional data analysis and allows complicated reaction pathways to be analyzed rigorously with no simplifying assumptions.

METHODS

Transient kinetic analysis in enzymology, which entails the study of events occurring at the enzyme active site, requires specialised instrumentation to measure reactions over the millisecond time scale. A *stopped flow instrument* allows enzyme and substrate to be rapidly mixed and the progress of reaction monitored by following an optical signal (e.g., absorbance or fluorescence). A computer is triggered to measure the intensity of light as a function of time after mixing for periods ranging from a few milliseconds to tens or hundreds of seconds. *A chemical quench flow* set up allows the mixing of enzyme and substrate to initiate a reaction followed, and after a defined time period, by mixing with a quenching agent (acid, base, or other denaturant) to prevent further reaction. The amount of product formed during the

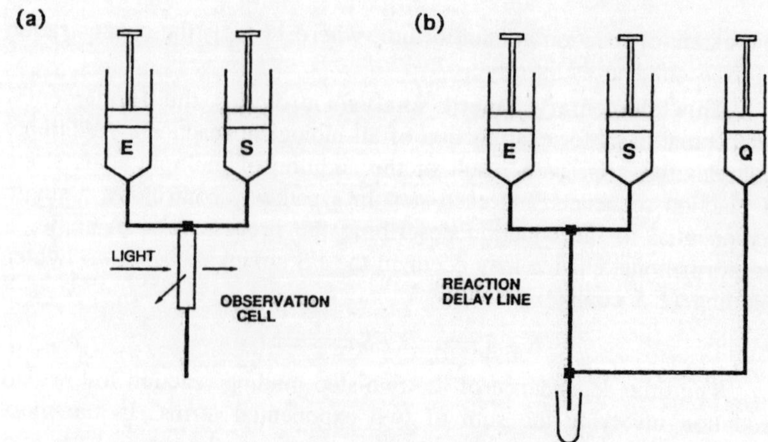

Figure 7.2 : The arrangment of syringes used to rapidly mix solutions in (a) stopped-flow and (b) chemical quench-flow instruments.

reaction must then be quantified by methods usually involving a chromatographic separation of reactants from products.

Modern stopped-flow and chemical quench-flow instruments employ a computer-controlled motor to drive syringes at a precise speed and distance to force the mixing of a defined volume of solution either into an observation cell, in the case of stoppedflow; or through a length of tubing before mixing with the quenching agent in a quench-flow instrument. The volumes of solution required for each reaction can be quite small (20-40 p.L) in the best of the modern instruments, opening up these methods to almost any enzyme system under investigation.

EXAMPLES

In the field of enzymology, transient state kinetic analysis has had a profound effect on our understanding of enzyme-catalyzed reactions. *Steady state kinetic* methods provide information pertaining only to the binding of substrates and to the magnitude of the rate-limiting step for catalysts in the form of k_{cat} and K_m. Information defining the events occurring after substrate binding and before product release is simply not available by analysis in the steady state. Transient kinetic analysis allows the direct measurement of events occurring during catalysis, in addition to providing direct quantification of binding steps. For example, transient state kinetic analysis has afforded resolution of enzyme-bound intermediates, the kinetics of ribozyme binding, and the mechanism of DNA polymerisation.

The enzyme EPSP synthetase provides the best example of the importance of a complete kinetic analysis of an enzyme reaction pathway by transient state kinetics. This enzyme, which is the target of the herbicide glyphosate, catalyzes the transfer of an enolpyruvoyl group from phosphoenol pyruvate to shikimate 3phosphate in a reaction that proceeds by an addition-elimination mechanism via a tetrahedral intermediate. It was the direct observation by rapid quench kinetic methods of the kinetics of formation and decay of the tetrahedral intermediate that led to its isolation and structure proof by NMR spectrometry. Estimates were obtained for the rate and equilibrium constants governing all six steps in the pathway.

Analysis of the mechanism of DNA polymerisation has been most profoundly affected by the application of transient kinetic methods. A comprehensive mechanistic analysis by rapid quench kinetic methods has shown that high fidelity of DNA replication is due to a two-step nucleotide-binding sequence. In the first step, the binding of a deoxynucleotide triphosphate (dNTP) to an enzyme-DNA complex is largely a function of ground state basepairing free energy and contributes a factor of 200 toward the overall fidelity in discriminating against an incorrect base pair. The rate-limiting step for polymerisation is a conformational change in the enzyme-DNA-dNTP complex which is a function of the basepairing geometry and contributes a factor of 2000 to the overall fidelity. The key steps in the reaction responsible for the extraordinary fidelity of DNA replication are transparent to steady state kinetics measurements, which are limited by the slow rate of release of DNA from the enzyme.

8

SECRETORY GRANULES

Knowledge that secreted proteins are synthesized in association with ribosomes found predominantly in the basal region of the acinar cell, are then stored in substantial quantities in apically placed *zymogen granules*, and are finally released across the cell's apical membrane into the duct lumen led to the conclusion that new digestive enzyme moves as a wave from its synthetic site of origin at the cell's basal pole to its apex. In this sense, the product can be said to follow a "*vectorial*" path. Although this anatomical polarisation is not always as striking as usually envisioned, it is nevertheless true that the great preponderance of ribosomes are found basally and the majority of secretion granules are seen at the apical pole, and of course the product eventually ends up, at least in substantial part, in the most apical site of all, the *duct lumen*. Thus, we can say with some assurance that as new protein travels through and out of the cell, we would initially expect to find it predominantly at its basal end, and over time at more apically placed sites, to and through the region of the zymogen granules, on its way out of the cell. Although, on the average, such movement of new protein appears to occur, "how" is another matter, and research efforts for over a quarter of a century have been directed at trying to find out in just what way or ways this transfer takes place. This is still a matter for considerable speculation.

FILLING OF THE GRANULES

All new protein enters the cisternae of the *rough endoplasmic reticulum* (RER) and that all cellular membranes are impermeable to these proteins once they have been so "*sequestered*." How would zymogen granules fill from this source? How would new digestive enzyme get from the internal spaces of the RER to the zymogen granules if the membranes of the RER are impermeable to them?

It is widely, although not universally, believed that the major subcellular membrane-bound compartments of the acinar cell are discontinuous. That is, the interior spaces of the RER are a separate compartment from the Golgi cisternae, which in turn are separate from the interior of secretion granules or condensing vacuoles, which are separate from each other as well. If this is the case, a special modus operandi would have to exist to move the newly sequestered protein from the compartment in which it finds itself (the RER) to its eventual storage repository (the zymogen granule). The cisternal packaging-exocytosis model proposes that "*vesicular*" processes account for this movement. Indeed, how else could movement occur? Vesicles that contain digestive enzymes are thought to be formed from the membranes of one compartment and their contents then transferred to another as a result of the fusion of the membrane of the vesicle and the membrane of the recipient compartment. In this way the movement of protein from the cisternal spaces of the RER to the zymogen granule would be achieved, and the absolute barrier that the membrane poses to the passage of protein broached.

What specific events have been proposed to account for this transfer? Unfortunately, the model, even now, some 25 years after it was first put forth, is not clear in this regard. This of course makes description difficult and experimental test regarding ambiguous points wholly problematic.

Vesicles from the RER

The removal of secretory protein from the cisternal spaces of the RER is thought to occur in small vesicles, often called *smooth-surfaced vesicles*, of about 400A internal diameter, formed, by budding, from the membranes of the RER. Budding is thought to take place at the end (apical end only?) of individual RER stacks at areas of the membrane that are denuded of ribosomes. The only evidence for such an event is the occasional appearance of narrowings at the poles of individual membrane stacks in electron micrographs that could reflect the beginnings of such a process. Such visual images are, of course, not very compelling evidence, as we have already discussed, and might well reflect other things, including static narrowings that have no functional significance whatsoever.

Even if we allow for the existence of budding, the proposal that protein is transported in this way still needs to be proven. That is, proof of budding does not prove that secretory protein is moved in this fashion, no less that such a process accounts for its movement quantitatively. In any event, there is no proof that such a process occurs beyond this evidence of form (narrowings), and the hypothesis that transfer occurs as the result of the budding of ends of RER stacks remains an essentially unsubstantiated hypothesis. There is no proof (1) of a dynamical kind that budding takes place; (2) that the small smooth-surfaced vesicles seen in these cells are derived from the membranes of the RER; or (3) that these vesicles in fact contain digestive enzymes.

COMPLEXITY OF THE RER

The complexity of an hypothesis has a great bearing on its adequacy as a theoretical framework upon which to develop an understanding of a system. While a system may seem complex to us as we approach it, we must nevertheless try to avoid complexity, to the degree that we are able, in the development of our hypotheses. As already discussed this among other reasons, is because a complex hypothesis, as evidenced by the number of subsidiary hypotheses that must be proposed, may be difficult, if not impossible, to prove false.

With this thought in mind, the budding hypothesis can be our point of departure for thinking about the considerable complexity of the cisternal packaging-exocytosis model. This single step or subsidiary hypothesis of the larger model requires the following subsidiary hypotheses of its own:

1. The existence of a specific force to produce the budding of the membranes of the RER.
2. A means of restricting expression of that force to certain locations on the membrane [only at the (apical) poles of RER stacks].
3. A means of excluding ribosome binding sites from budding sites.
4. The replacement of membrane at a rate equal to the rate of budding so that at the steady-state the amount of RER membrane is constant.
5. And, of course, to produce the above effects, the existence of parts of RER membrane specifically (chemically) modified to carry them out.

There are other subsidiary hypotheses that we could add, such as a mechanism for selecting those intracisternal proteins destined for secretion granules from others that are not. This list is not given to be complete, but to point out that what may seem a simple proposal at first glance, the budding of membranes, is really quite a complex idea, at least in this particular setting. The level of complexity becomes even more profound when we consider further details, such as what additional processes and mechanisms are required to account for the replacement of membrane at a rate equal to the rate of budding. The important thing to keep in mind here is that these complex ideas are not being considered because our knowledge of the kinetics and dynamics of the process requires such complexity. That would be justifiable. But rather, because our *assumption* that membranes are impermeable to proteins requires it.

MATHEMATIC SOLUTION

To some the absence of evidence for budding may represent something of a challenge. That is, they would seek the missing evidence. And if they were successful in finding something in the parts of the cell or its behavior that is consistent with this hypothesis, might conclude that they have provided evidence in support of the idea. And such evidence it might well be, but without knowledge of how it fits with an actual kinetic and thermodynamic description of an intact, known process of *budding*, it might be fragile evidence indeed. In this light and before moving on to the next step in the model, we should consider the complexity of the budding hypothesis from a dynamic point of view; that is, what would be required of such an event if it accounted for the transfer of enzyme, based on what we know of secretion.

Many of the calculations that follow merely reflect the geometric fact that as a sphere becomes smaller the ratio of surface area to volume becomes increasingly large, and hence the smaller the vesicle, the smaller the mass carried per unit diameter or per unit membrane surface area. The vesicles that are thought to be formed from the RER are of about 400 Å internal diameter, and therefore have an internal volume of some 3×10^{-17} ml. How many molecules of secretory protein would each such vesicle carry? The proteins in the zymogen granules are predominately digestive enzymes, and their concentration there has been estimated for the rat and rabbit to be between 4-9 mM for an average mixture. The concentration of secretory protein in the cisternal spaces of the RER can be estimated by comparing the specific activity of digestive enzymes in the RER (in microsomal fractions of tissue homogenates, in point of fact) to that in the zymogen granule. For example, microsomal specific activity is about one-fifth of that in the zymogen granule in the rat, and therefore, the concentration of these proteins in the RER can be estimated at about one-fifth of that in the zymogen granules or about 0.8-1.6 *mM*. Much of the enzyme in microsomes appears to be adsorbed to the membrane or associated with the ribosomes and thus a better, but still conservative, estimate would be about half this value. By this calculation, if the concentration is the same in the small vesicle as in the cisternae, then each vesicle would contain 1.3-2.6 $\times$ 10^{-23} moles or approximately 8-16 protein molecules per vesicle on the average. For ease of calculation, let us assume that there are 10 secretory protein molecules within each formed small vesicle.

At a steady-state in which protein synthesis and secretion are equal, the rate of *vesicle formation* (that is, the number of vesicles X the content of each vesicle) would equal the rate of protein secretion. If this balance or steady-state can be maintained at moderate rates of *protein secretion*, let us say 25% of a maximal rate of about 10-20 mg/hr/g of tissue (based on observed rates in rat and rabbit), then about 3-7 $\times$ 10^{-7} moles or 1.9-4.4 $\times$ 10^{17} molecules of protein (using an estimated average molecular weight for the secreted digestive enzymes of 30,000 daltons) would have to be packaged in small vesicles each hour per gram of tissue. From this we can calculate the number of vesicles containing 10 protein molecules apiece that would have to be formed per unit time. For each cell (with a volume averaging about 1000 μm^3 or 1 $\times$ 10^{-9} ml), 1.9-4.4 $\times$ 10^8 molecules would be synthesized and secreted per hour on the average. At 10 molecules per vesicle, 20-40 million vesicles would have to be formed per cell per hour. This would require an amount of membrane equal to 170-330 times the surface area of the cell's plasma membrane (using cell surface area = 600 μm^2 and small vesicle surface area = 5 $\times$ 10^{-3} μm^2). However one might quibble with the particular numbers a remarkably active process of membrane turnover or recycling, to my knowledge uniquely so among cells.

We can also ask a related visual question. If we assume that the narrowings observed in RER sections reflect an event premonitory to budding, then how frequently should we see such forms given a rate of 2040 million vesicles being formed per hour per cell. Converting to seconds, this would be approximately 5,000-10,000 vesicles/cell/sec, or an equal number of narrowings in every cell each *second*. Does the observed frequency of narrowings agree with this calculation?

If we estimate that substantial amounts of RER would be seen in 75% of 0.08-μm-thick electron microscopic sections through a 10-μm-thick cell, or in 94 sections through each cell, then we would expect to see about 1/100th of the total number of narrowings in each RER-containing cell section or about 50-100 narrowings or forming vesicles in each and every section if such an event was of the order of one second in duration. This is clearly not observed.

One might argue, as has been regarding exocytosis, that the process occurs too rapidly for us to be able to observe more than a small fraction of the total number of events. For example, for a millisecond event, a very fast biological occurrence of about 10-20% of the duration of a fast action potential, assuming a proportional reduction from a 1-sec event, only 5-10 narrowings would be seen in each cell. Even this frequency may be greater than what is actually observed, although it has not been quantified.* Of course, we could posit an even faster process if necessary. With constructed hypotheses, such as the budding hypothesis, a low frequency of a form can be taken as evidence of the rapidity of the process, without any real knowledge of its existence, no less its rate.

SMOOTH-SURFACED VESICLES

Despite these problems, let us assume that budding occurs and accounts for the transfer of protein out of the RER quantitatively. Next we would ask, how do these vesicles move? The model proposes that they "*migrate*" to the Golgi region of the cell where they empty their contents into the lumena of waiting condensing vacuoles and perhaps into Golgi cisternae as well (N.B.: Certain sugars are thought to be added to sugar-containing secretory proteins in the *Golgi complex*. If this occurs within the Golgi cisterns, then we must either propose vesicle budding at this locus or that condensing vacuoles are formed from these structures).

If the evidence for budding is weak, then it is nonexistent for the movement of vesicles. There is no experimental evidence that the small vesicles seen in the acinar cell even exhibit *Brownian motion*, although one assumes that they do, no less directed movement, since they have only been seen statically under the *electron microscope*. Indeed, with some exceptions, there is little evidence for the directed movement of granules or vesicles thought to be involved in secretion processes in general. Sometimes differences observed in their location in the cell from time to time or situation to situation are offered as proof of movement. For example, when there are fewer *zymogen granules* in the acinar cell, they tend to be apically placed, and the inference may be drawn that they have moved to the apical surface from other locations. However, from electron microscopic images of fixed or otherwise nonliving tissue, movement cannot be observed, only imagined. The fact that the secretory product moves is also sometimes viewed as being tantamount to proof that vesicles, thought to contain that product, also move. Of course, even assuming that one can demonstrate that the latter is contained in the former, the movement of the vesicle and its contents are separate matters, unless one can show that the contents cannot leave the vesicle and move independently.

None of this is to say that vesicles do not move, but rather that direct evidence for movement has been lacking. Moreover, even if we could demonstrate movement, it alone would certainly not validate a vesicular transport hypothesis. We would also have to demonstrate that vesicle movement is directed and occurs at a rate that is quantitatively appropriate for the process. It is a particularly telling fact I believe that there is no convincing evidence that vesicles move in situ, no less in a directed and quantitatively appropriate manner, when such movement is central to the definition of *vesicular transport* hypotheses—that is, the directed movement of vesicles acts to transport product from place to place. Given this fact, I find the spate of studies describing the role of intracellular tubules and filaments in such movement premature. Perhaps the hope is that once the details of the mechanism become clear, so will the existence of the process it carries out.

CONDENSING VACUOLE

Let us put aside the problems with a vesicular mechanism up to this point and assume that the smooth-surfaced vesicles do contain digestive enzymes and somehow reach the *condensing vacuoles* at a satisfactory rate, fuse with them, and release their contents within the vacuolar lumen. Now new problems arise. We must assume that this fusion is a specific event, that is that specific membrane structures exist to promote fusion between these particular membranes, as well as prohibit fusion between the vesicles and undesirable cell components; for example, small vesicles should not fuse with each other, nor condensing vacuoles with each other, nor either with fully formed *zymogen granules*, nor should any of these fuse with mitochondria, at least not to an appreciable degree. Compared with some o! the complexity that we have already encountered, this may seem to be a rather insubstantial problem. Lipid bilayers certainly seem capable of fusing with each other, and a couple of proteins might well take care of the issue of specificity.

Nevertheless, however simple or complex one may think these ideas, there is no significant body of evidence, either anatomic or dynamic, describing such events. The fusion of smooth-surfaced vesicles to condensing vacuoles is, at the least, an uncommon image in electron micro-graphs. We can of course again argue that the events occur too rapidly to be seen at a substantial frequency.

DYNAMICS OF FUSION

Now let us take a more detailed look at the idea that small vesicles fuse with the condensing vacuole. What would we expect to see if such an event occurred? We can use our fact about secretion, a known rate of protein secretion, for help. We have already calculated that about 5,000-10,000 vesicles should be formed per cell per second in order to account for observed rates of secretion and thus an equal number of fusion events should occur at the condensing vacuole each second. Thus, vesicles formed from the extensive membranes of the RER, occupying some 20% of cell volume, now converge on a minor cell pool, the condensing vacuole which occupies about 2% of the volume of the average cell. From stereological measurements we can estimate that the total volume of the condensing vacuole pool is about 15 μm^3. If we consider the average condensing vacuole to be a little less than 1 μm in diameter or about 0.5 μm^3 in volume, then we would have approximately 30 condensing vacuoles per cell. As I have said, small vesicles would fuse with the *condensing vacuoles* at a rate of about 5,000-10,000/sec, or for a population of 30 condensing vacuoles, about 200-400 fusion events per second per vacuole. And yet such events are rarely, if ever seen. Even if we assume a very fast process, such images should still be seen regularly, and to my knowledge they are not.

Perhaps what would be required can be appreciated in another way. If the completely filled condensing vacuole (now a *zymogen granule*) contains protein at an average concentration of 5 mM and if the average granule volume is about 0.5 μm^3 (for granules of about 1 μm average diameter), then each zymogen granule would contain about 1.5×10^6 protein molecules. If 10 molecules were deposited at the condensing vacuole per small vesicle fusion, then 150,000 such events would be required to fill each granule. The surface area of one small vesicle is about 5×10^{-3} μm^2, and thus, about 750 μm^2

of membrane or an amount of membrane about equal to that enclosing the cell would be required to fill each and every zymogen granule.

CONDENSING VACUOLE

But our problems with the *condensing vacuole* are really only beginning. The distribution of condensing vacuole size is quite similar to that seen for the "*mature*" zymogen granule population. That is, partially, even virtually empty granules, are seen over a broad range of sizes from those equivalent to the smallest zymogen granule to those as big as the largest. Thus, one often sees large condensing vacuoles that contain very little material, and that would presumably not be enlarged further as digestive enzyme is added. The conclusion that is drawn from this observation is that condensing vacuoles do not grow from small vesicles to larger ones as they fill, but that granules of a range of fixed sizes fill. And thus according to a vesicle model, all 150,000 small vesicles must bud from the condensing vacuole after having deposited their load.* If budding at the RER is a complex process, it is no less so at the condensing vacuole. As with small vesicle fusion, the budding of vesicles from condensing vacuoles does not appear to be a regular feature of electron micrographs of these structures, and there is no other evidence that such events occur.

1. A force to initiate budding.
2. And specific (chemical) membrane components to carry it out.
3. In addition, a means of leaving the transported content behind, that is, some way of not reincorporating the material just released back into the budding vesicle.
4. A means of incorporating other contents of the condensing vacuole to fill the space.
5. Since there are large differences in the surface area to volume ratio for the two vesicles (about 6:1 for the condensing vacuole and 200:1 for the small vesicles), there would be a "shortfall" in the amount of material transferred as a result of fusion, that is the volume of the larger vesicle would become larger than the space filled by the contents of the smaller vesicle (it would fill 1/30th of the newly formed space), and the opposite situation would apply due to budding, that is, too much membrane would be removed (30 times too much relative to volume). Thus, budding and

fusion would either have to be coupled to deal with these discrepancies or fluid would have to enter and exit the condensing vacuole in order to make the necessary volume adjustments.

6. If the condensing vacuoles are of relatively constant size, then to maintain size stationary, the rate of fusion of small vesicles and their subsequent budding would in any event have to be equal, that is, there would have to be a means of regulating the rates of the two processes relative to each other (as well as relative to the rate of budding and fusion at the RER).
7. Once formed such "*empty*" vesicles would have to migrate in a directed fashion back to the RER and be reincorporated therein, in this way replacing RER membrane lost by budding.
8. There would have to be a means of distinguishing between the transfer of empty vesicles to the RER and the transfer of filled vesicles in the other direction.
9. It would have to be unlikely for empty vesicles to fuse with condensing vacuoles and equally unlikely for filled vesicles to fuse with RER.
10. And finally the "structure" of the two vesicles would have to be different in order for this to occur, and therefore membrane components would have to be exchanged in some fashion at both the condensing vacuole and RER.

Whether or not such things are conceivable, the point is merely that such a proposal is very complex and is virtually unsupported by direct experimental evidence of any kind.

FILLING OF CONDENSING VACUOLES

The membrane-bound compartments within the secretory cell are thought to be separated from each other, that is, are discontinuous, and that this view, although widely held, was not universal. It has been suggested that the spaces of the RER and the Golgi apparatus are connected and therefore that material can be moved within the cisterns rather than between them. The anatomical evidence offers convincing support for this view, such a system would help us divest ourselves of the complex system of small vesicles. In a continuous system, protein within the RER would merely diffuse, or otherwise move, within these internal spaces of the cell to the region in which condensing vacuoles are formed.

Although this idea seems far superior to the vesicle shuttle concept because of its relative simplicity, its superiority unfortunately ends at the Golgi. For even if RER cisterns are continuous with the spaces of the *Golgi complex*, and moreover, even if condensing vacuoles are formed from Golgi membranes, it seems that large, relatively empty condensing vacuoles exist as free-standing entities, not connected to the *Golgi membrane* or any other cell structures. If this is true, and if we believe that the granule membrane is impermeable to protein, then this barrier must be broached by a discontinuous vesicular process in order for granules to fill, and we are thus back where we started.

MEMBRANE FLOW

When it was first proposed that new protein is moved from the *ribosome* to the *zymogen granule* by vesicles, it was thought that transfer was accomplished as a result of the flow of membrane from compartment to compartment; RER membrane became Golgi membrane became zymogen granule membrane, and so forth. However, the result of over 25 years of research has been essentially negative in this regard, at least for the acinar cell, and the expected mixing of membrane constituents has not been demonstrated. The protein and lipid content of the various cellular membranes thus far seem to be unique and particular. This apparent failure of a general membrane flow theory is partly responsible for the development of the "*shuttle*" concept that we have just discussed.

DISAPPEARING ZYMOGEN GRANULES

A decrease in the number of *zymogen granules* in the acinar cell during augmented protein secretion was first noted by Rudolf Heidenhain in the 19th century, and has been reported many times since. To the degree that such a decrease is a prediction of the exocytosis model, it is borne out by the evidence. Indeed, this, perhaps more than any other observation, is responsible for the widely held belief that exocytosis occurs and accounts for secretion, because such a process could account for the disappearance of granules during active secretion.

Constancy of Number

Before we discuss the disappearance of secretion granules, we should consider the possibility that granule number may not decrease during augmented secretion. How would such an eventuality affect our consideration of the exocytosis hypothesis? In the first place, it is often unclear when a reduction in number has or has not occurred. In Heidenhain's original studies on dog pancreas, as in most subsequent light microscopy studies, the term granule "*depletion*" is used to indicate that the area of the cell occupied by zymogen granules, the granular zone, has become smaller. This does not necessarily mean that the number of granules in the cell has decreased. Of course, the same number of granules might merely occupy a smaller volume of cytoplasm, and this could occur for a variety of reasons, one of which (a decrease in their size). This distinction is often unappreciated, and decreases in the number of granules in the cell has become the common explanation for decreases in the size of the granular zone whether or not such an inference is actually justified by the data. Most reports concerning changes in granule number have not been quantitative, and, as a result, often reflect little more than the impressions of the microscopist, however carefully made. *Quantitative measurements* from which we can more objectively estimate granule number in various functional states are needed and necessary.

Recently, Ermak made one such estimate. He found that although *active secretion* (induced by feeding) may lead to a reduction in granule number, the number of granules in some individual cells does not decrease. That is, active secretion does not *require a* reduction in the number of granules in cells.

At first glance, this appears to contradict the paradigm's prediction that granules should disappear under these circumstances. There are, however, two obvious ways in which we can account for the absence of changes in granule number during active secretion within the context of an exocytosis mechanism. The first is that if the number of granules in a particular cell or group of cells does not decrease during augmented secretion, then these cells may be refractory; that is, they may not be secreting at all, or may only be secreting at relatively low rates. In order to test this hypothesis properly it is necessary to compare rates of protein secretion for different individual cells, relative to changes in the number of granules in the same cells, and such measurements are not yet practicable. Ermak has however reported that after a period of augmented secretion, changes in the *size* of granules are found in cells in which granule number is unchanged. This is a "*cytologic*" sign of activity, and although in itself it does not prove that such cells are actively secreting protein, it does indicate that they appear, at least in this

Table 8.1 : Mean Diameter, Volume, and Number of Zymogen Granules in Acinar Cells of Fasted and Fed Rats

Experimental Group	Mean diameter (μm)	Calculated volume (μm^3)	Mean number
Fasted	0.85 ± 0.15	0.32	45 ± 3
	= 916		n = 54
Fed	0.65 ± 0.15[δ]	0.14	29 ± 288
	=1660		n=119
Sample *1C*	0.70 ± 0.20[δ]	0.18	43 ± 4
	= 256		n = 12
Sample 10	0.60 ± 0.15[δ]	0.11	15 ± 388
	= 140		n = 15

fashion, responsive to the applied stimulus. That protein secretion is indeed occurring in such cells can be inferred from the extent and nature of the observed changes in granule size, and this will be discussed below.

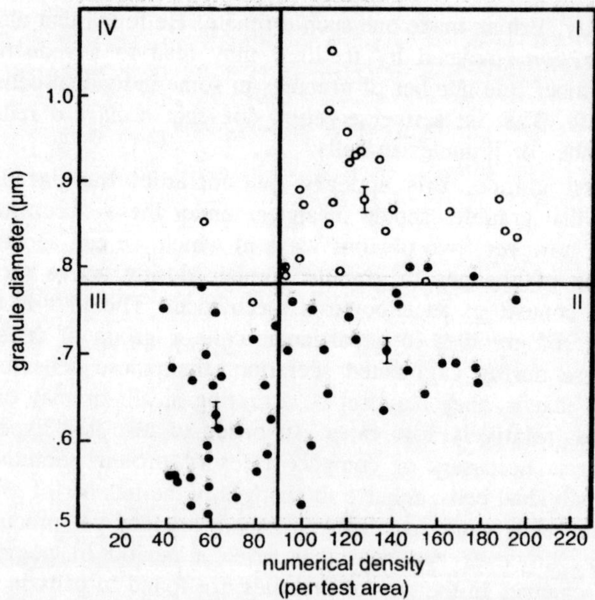

Figure 8.1 : Scattergram of the mean diameter of granules versus numerical density (the number of granules per test area) (test area approximately 275 μm^2) from fasted (O) and fed (•) rats.

The second explanation is in a sense the opposite of the first; that is, the rate of new granule formation in some cells is so active that granule number remains constant despite a large increase in the rate of secretion. That is to say, cells that do not show changes in granule number *are more active*. However, for a number of reasons it does not seem likely, at least in the pancreas, that the rate of granule formation can keep pace with active secretion. From cases in which large decreases in granule number are observed during augmented secretion, we can estimate the rate of granule formation under these circumstances for the tissue overall, at least as we consider these questions through the eyes of the paradigm. For example, in one study, the granule content of the cell was reduced by some 95% over a 3-hr period of *augmented secretion*. For this to have occurred, the rate of granule formation would have been only 5% of the rate of loss in the paradigmatic view of the process.

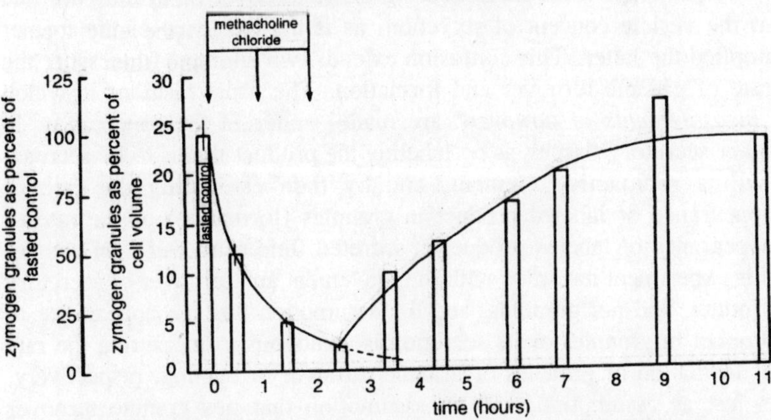

Figure 8.2 : The effect of a cholinergic agonist (metacholine chloride) on the percentage of cell volume occupied by zymogen granules as determined by a random point-count method in light micrographs.

Of course, the point is that their rate of formation may be higher in the particular cells or tissues in which depletion is not seen, but it should be kept in mind that the rate of granule formation in such cells, given an equivalent rate of protein secretion, would have to be much greater than in cells in which substantial depletion is observed; more than 20 times as great for this particular example. Moreover, under conditions in which protein secretion is maximally increased (by 10-100 times), protein synthesis is hardly altered in relative terms (a brief increase that is maximally about two- to threefold for most reports). To the degree that protein synthesis reflects the rate of granule formation, it does not appear able to keep pace with active secretion. In any event, potentially one can account for the absence of changes in granule number during active secretion within the context of the exocytosis model by means of these ad hoc hypotheses, namely, increased granule formation or decreased secretion. However, it should be remembered that these explanations are simply conjectures, that is, unsupported by evidence.

GRANULE TURNOVER

It is believed that granules are formed continuously and that their formation accounts for thc increase in the number of granules in the cell during the period of recovery following active secretion, and furthermore, a decrease in their number during active secretion is thought to occur because the rate of granule formation is less than the rate at which granules are lost or disappear.

Sometimes these ideas about granule turnover or cycling are tied to the vesicle concept of secretion, as if the existence of the former implied the latter. This confusion extends two attempts to measure the rate of granule turnover and formation. The usual fashion, in which "*measurements of turnover*" are made, either in the pancreas or in other secretory tissues, is by labeling the product that is to be secreted with a radioactive element, and by then estimating the rate of appearance of labeled product in granules (formation) or the rate of appearance of labeled product in secreted fluid (turnover). Of course, this experiment has to do with the movement and turnover of secretory product, and not granules, and the assumption that the appearance of product in granules or its secretion is analogous to measuring the rate of formation of granules or granule turnover or cycling, respectively, is just as assumption; and an assumption that ties granule turnover arbitrarily to the vesicular concept of secretion. That is, it assumes a priori that the turnover of granules and product are one in the same and that the secreted product cannot and does not move from place to place independently of the formation, movement, and loss of granules. This assumption cannot be made. Indeed, it is known, not only for the pancreatic zymogen granule, but for many secretion granules, that the secreted contents enter and leave the granule across its membrane, as well as enter and leave the cell across the plasma membrane; and therefore the two processes are often known to be independent.

It is an important test of the *exocytosis hypothesis* to determine whether or not the rate of appearance of product in granules and the rate of turnover of product are indeed the same as the rate of formation and turnover (or cycling) of granules. If they are the same, then this would be support for the exocytosis concept. Nevertheless, we have no right to assume the equality of granule and product turnover a priori. To measure the rate of granule formation and turnover we must have a specific granule marker that is *independent of the secreted product,* perhaps a membrane marker (N.B.: Meldolesi has reported that the rate of turnover of granule membrane protein is lower than that of secretory protein).

Although it is true that without a granule marker we cannot properly measure the rate of granule turnover, we can nevertheless ask a simple question that can help us evaluate the situation and set limits in the absence of this measurement. How much of the product present in the cell at the outset of augmented secretion is secreted during an active period? Is it several times that originally present in the tissue or is it less? If it is the former, then at least we know that the product, if not the granules, has turned over numerous times during the secretory period, and this would at least give us cause to consider the role of granule turnover under these circumstances. On the other hand, if the amount that is secreted is only a fraction of the product present in the granule pool or the cell at the outset of augmented secretion, then substantial turnover of product would not be necessary to account for secretion, and in its absence, in paradigmatic terms, neither could substantial granule turnover or cycling have occurred.

If the latter were indeed the case, then it would considerably simplify our attempts to understand the nature of changes in granule number, or the absence of such changes, and their relationship to protein secretion. What are the facts? In the fasted rat pancreas, the amount of secretory protein in the cell is about 70 mg/g of tissue. Depending upon our assumptions, the granule pool would contain between 30 and 60 mg of protein/g of tissue. Three hours of maximal stimulation with a cholinergic agonist in the rat causes secretion of about 20 mg of protein/g of tissue. Therefore, only a fraction, between one-third and two-thirds, of the original granule pool of secretory protein is released from the cell even during a prolonged period of active secretion. Thus, the enzyme pool does not turn over numerous times in order to account for active enzyme secretion. The initial store is sufficient.

In the context of the *exocytosis model*, if this secretion is not accompanied by a change in the tissue content of secretory protein, then new granules presumably would have been formed at a rate equal to their rate of loss. On the other hand, if tissue content decreased in direct proportion to the secretion of protein, then no new granules would have been formed (assuming unchanged intragranular protein concentration). When such measurements were made for the above experimental situation, the observed decreases in tissue content ranged between one-third and onehalf, suggesting that the latter view, that is, the rate of new granule formation is low relative to their loss during active secretion, is closer to the truth.

In this case, we would also expect to find a reduction in the volume of the cell occupied by granules (*granule volume density*) of one-third to twothirds under the same circumstances. When quantitative estimates were made, the reduction was even larger, about 95%, reinforcing the view that the rate of new granule formation is low relative to their loss under these circumstances. This decline in the granule content of cells can be used to estimate the rate of granule formation. When estimated in this way, a value is obtained of about 1.5% of the initial pool being formed per hour. We can also estimate the rate of granule formation from the rate at which granules reappear during recovery from stimulation. In this particular case, full recovery is observed in 6-7 hr or at an average rate of about 14-16% per hour.

Whether the estimate is 1 or 15%, the rate of new granule formation cannot be very large relative to the size of the granule pool found initially in the pancreas of the fasted rat, and changes that are observed in the granule pool over a short period can be considered to involve, with some small error, the same granules that were present at the outset, even during the most active secretion. It should be kept in mind that these estimates assume the exocytosis model for secretion, and in other theoretical contexts the true rate of granule turnover might be quite unrelated to these observations.

DISAPPEARANCE OF GRANULES AND SECRETION OF PROTEIN

Whenever there are decreases in the granule volume of a cell the rate of digestive enzyme secretion is increased, this decrease to be correlated to the amount of protein secreted. The paradigm is quite specific in this regard.

It is generally believed that the concentration of digestive enzyme protein is either essentially the same in each zymogen granule, or varies randomly from *zymogen granule* to *zymogen granule*. As a result, the amount of enzyme stored in a granule, or more generally in the granule pool of the cell, can be said to be roughly proportional to its volume, or to the fraction of the cell volume that it occupies, its volume density. Therefore, if granule sizes are normally distributed, as they more or less are, then the product of the number of granules in a cell and the average volume of the individual granule is a measure of the amount of enzyme stored in the granule pool of that cell.

In this case, changes in granule volume density should be inversely proportional in a linear fashion to changes in protein secretion, as long as the formation of new granules is slow in comparison to the rate at which the contents of formed granules are secreted. That is, if secretion is accomplished by the direct transfer of mass from one compartment (*intracellular granules*) to another (the *duct lumen*), then changes in the content of the two compartments should be reciprocal and proportional; as more and more enzyme is secreted, fewer and fewer granules should remain in the cell, and volume density should decrease accordingly. Eventually the rate of secretion should decrease to zero or slightly above zero (to a rate at which granule formation equals the rate at which protein is secreted) as the cell becomes depleted of its granule contents. If we plot the decrease in granule number against the amount of protein secreted, a linear function should be formed that intercepts the axes at approximately the origin (no loss of granules, no secretion of enzyme), and in addition, our function should reach a limit at 100% of the granules lost (no more can be secreted than can be accounted for by 100% of the available granules).

Therefore, in terms of the exocytosis model, during the augmented secretion of digestive enzyme: (1) decreases in granule volume density should be proportional to decreases in granule number; (2) decreases in granule number and volume density should be inversely proportional to increases in the rate of protein secretion; and (3) augmented secretion should cease when the cell is depleted of its granules.

GRANULE NUMBERS AND VOLUME DENSITY

As already mentioned, decreases in the size of granules are seen during active secretion, even in the absence of changes in number. In this case, *all* of the decrease in granule volume density is attributable to changes in the *size* and not the number of granules. Changes in granule size during *augmented secretion* were first noted by Heidenhain's contemporaries, Kiihne and Lea, viewing living rabbit pancreas in the light microscope, and were confirmed by Heidenhain in fixed dog *pancreas*. While the changes in the number of granules in cells during active secretion that were reported by Heidenhain formed the foundation of the vesicular theory of secretion, the roughly contemporaneous observation that granules change size has lived a life of complete obscurity until quite recently. The literature contains numerous, and sometimes rather striking cases, where subsequent to the application of a secretory stimulant, the size of granules in cells decrease. Often the fact is not noted, and perhaps not even noticed. Or if noticed, assumed to be due to the presence of newly formed granules. Recently, Ermak undertook a reexamination of Kiihne and Lea's observations of size change in the pancreas under the electron microscope. He measured changes in the size and number of granules in cells by feeding rats whose acinar cells were filled with zymogen granules after an overnight fast. He found that large changes occurred in the size of granules and that the average size decreased by over 50% after 90 min of feeding. These decreases were seen in cells that appeared to have fewer granules, but also in those in which no change in granule number was seen. Moreover, changes in granule number were not seen without accompanying changes in granule size. This is important because it suggests that changes in size and number are related. Beyond that, it suggests that a decrease in granule number may *depend upon* a decrease in granule size; that is, it may be that if and only if granule size decreases, does number decrease.

This is also consistent with the fact that smaller granules are sampled less frequently in electron micrographs, and therefore lead to an underestimate of true granule number relative to larger granules. It is often overlooked that substantial apparent decreases in granule number may merely be illusory, due to differences in sampling probability when granule size is reduced. Therefore, granule disappearance, real or illusory, may occur *because* size decreases.

Thus, the prediction of the paradigm that changes in volume density are the result of changes in granule number is clearly incorrect, and changes in size can account for a large percentage, and under some circumstances all, of an observed change in volume density during active secretion; and most significantly, decreases in number may even be attributable to changes in size.

Decreases in granule size can be explained within the paradigm in two ways: (1) decreases in size reflect the formation of new, smaller granules, and (2) the decrease in size does not indicate the loss of protein from the granule but is attributable to an "*osmotic*" event. In the former case, for the feeding experiment just described, we would have to propose that during the 90-min period of feeding, the granule pool had turned over numerous times. If the pool had been replaced just once in 90 min, then some 50% of the granules originally present would still be in the cell and of course be of the original size, thereby producing a bimodal distribution of granule sizes. This is not seen. The distribution of granule sizes after feeding is essentially the same unimodal log-normal distribution observed in the fasted animal, but displaced to lower mean granule diameters.

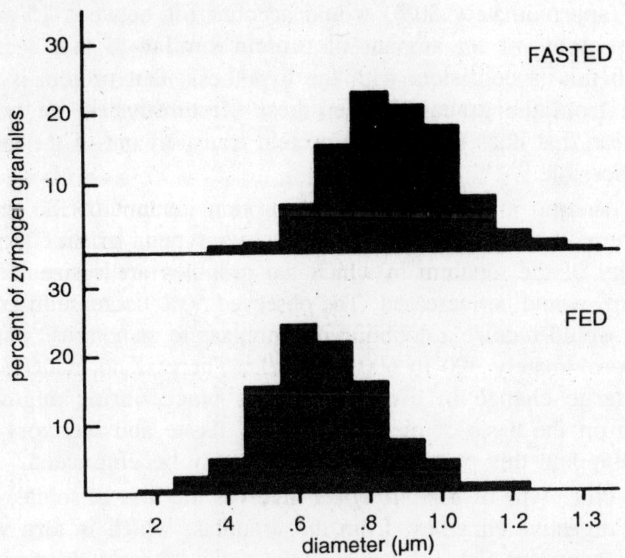

Figure 8.3 : Distribution of the diameter of zymogen granules in the pancreas of fasted and fed rats. Data expressed as percentage of total population.

A more complex notion can also be considered, that is, old granules could be secreted first, new ones following in line. This hypothesis would only require that the granule contents of the cell turn over once during the experimental period in order to produce a new normal distribution of granule sizes. However, there is no indication in micrographs of these tissues that large granules are lined up at the apical surface of the cell, while smaller ones are placed more basally. The granules appear to be more or less randomly distributed in terms of their size.

In any event, from our previous discussion, it seems clear that only a small fraction of the granule pool could have been replaced during the 90min period of feeding, and therefore the decrease in the size of granules, some 50% on the average, was an occurrence that primarily involved the same granules that were present at the outset of the study; that is, individual granules decreased in size during active secretion.

We can estimate the amount of protein secreted into the duct system during this 90-min feeding period maximally at about 15 mg/g of tissue. If we estimate the amount of enzyme that would have been lost from granules, if release from them were proportional to the changes in granule *volume*, then the observed change in granule volume (approximately 50%) would account for between 15 and 30 mg of protein, or an amount of protein similar to that secreted. Although this is consistent with the hypothesis that protein is being released from the granules under these circumstances, an osmotic explanation that does not involve protein transport out of the granule is also possible.

An osmotic effect in which the protein content of the granule would remain unchanged could be of two types. In the first, the osmolarity of the medium in which the granules are suspended, the cytoplasm, would be increased. The observed 50% decrease in granule volume, would require a doubling of cytoplasmic osmolarity, namely, from approximately 300 to 600 mOsm/L. There is no evidence that such a large change in osmolarity takes place during augmented secretion on the basis of measurements of tissue and secretory fluid osmolarity, and this possibility can reasonably be eliminated.

The other type of *osmotic effect* involves the loss of solute, other than the digestive enzymes, from the granules, which in turn would cause a dependent flow of water in its wake, thereby leading to a decrease in granule size. For such an effect to account for substantial changes in granule size, the solute being moved must be the major solute of the intragranular medium, that is, the solute that accounts in great part for its osmolarity. In addition, there would have to be a concentration gradient for this solute during unstimulated secretion, because its dissipation would be required to provide the force for the efflux of water. A twofold concentration gradient could account for a 50% reduction in granule volume. Potassium salts would seem to be the most likely candidate based on their high concentration in tissue, however, the membrane of the zymogen granule does not contain a Na^+, K^+-ATPase thought necessary for the maintenance of a concentration gradient for K^+.

Whatever gradient-producing mechanism we might consider, an

osmotic effect seems unlikely for another reason. In the feeding experiment just described, the change in granule volume was 50%, but only on *the average*. Since volume is a cubic function of radius, a given change in the radius of a large granule would produce a smaller relative decrease in volume than an equivalent change in the radius of a small granule. Since roughly equivalent changes in the radius of granules of various sizes is what occurs (as evidenced by a shift in a normal distribution of granule diameters), the same percentage change in volume does not apply to each granule. Indeed, because we are dealing with a cubic function, the range of differences can be immense. An osmotic effect requires that the changes in volume be proportional; that is, that the volume of large, medium, and small granules all change by the same proportionate amount—about 50% in the study cited. We can appreciate the very great deviation from this prediction by considering the case of the small granule. If we consider conservatively that the smallest granules seen after stimulation were derived from the smallest present before stimulation, then very large changes in the volume of these granules, in relation to the average change, must have taken place. For example a change from a 0 4-μm diameter granule to one of 0 1 μm the approximate lower limits in both fasted and fed groups, would result in a decrease in granule volume of more than 99%. Far greater than the average or mean value of 50%!

This example also shows why the size changes must include the loss of protein from the granule. Estimates of the contents of the *zymogen granule* indicate that about 60% of its mass is water and about 40% solids, in great part the digestive enzymes themselves. Thus, an osmotic effect could maximally reduce the volume of the granule by 60%, if it were totally depleted of intragranular water. Thus any change in volume greater than 60% would require the loss of solids (protein) contained in the granule. For the smallest granule case considered above, virtually all its protein contents would have been released (over 99%).

Thus, the prediction of the paradigm that changes in the volume of the cell occupied by zymogen granules, that is, changes in their volume density, are due to and are therefore strictly proportional to changes in granule number is clearly incorrect. Granule size also changes with changing functional state, and not only must be factored into the equation as a cause of changing volume density, but almost wholly accounts for such changes in some situations. Certainly, the "*depletion*" of granules and reductions in the volume of the cell that they occupy cannot be assumed to solely reflect decreases in their number.

During the prenatal period when granules first form in the acinar cell, over 90% of the increase in their volume density is attributable to changes in granule *volume,* and not number. Similarly, immediately after birth when active secretion first starts, granule volume decreases by about an order of magnitude during the first days and weeks of life. Thus, the size of granules varies with secretory state, and a "*granule size*" secretory cycle exists that is analogous to the classical "*granule number*" secretory cycle in which granule number decreases with feeding and increases after digestion is complete. That is, it also appears to be true that granule size decreases during active secretion and increases during periods of inactivity and increased storage.

GRANULE VOLUME AND THE AMOUNT OF ENZYME SECRETED

The paradigm predicts that the loss of granules should be reflected in a proportional increase in the amount of protein secreted by the gland, that is, it predicts that these two variables are inversely proportional to each other in a linear fashion. Some years ago we tested this prediction in rats, measuring both the volume density and the amount of enzyme released during highly augmented secretion. When changes in the two variables were plotted against each other, the relationship was found to be nonlinear. The ratio of granules lost to protein secreted was greatest at the onset of augmented secretion when the gland was filled with granules, and it decreased as the number of granules in the tissue diminished. Thus, the relationship between the change in volume density (number of granules lost in the paradigm's terms) and the amount of protein secreted was variable. It was not fixed or constant in accordance with the exocytosis model. Less protein was secreted for a given amount of "*granule loss*" when the cell was filled with granules, than when granule volume density was low.

To explain this nonlinear relationship in terms of exocytosis, we must propose that the rate of granule formation increases during augmented secretion, becoming increasingly greater as fewer and fewer granules remain. In this way it would only appcar as if fewer granules accounted for a given amount of protein secretion, whereas in fact the rate of cycling of granules had been increased instead. Such an increase in the rate of granule formation would have to be

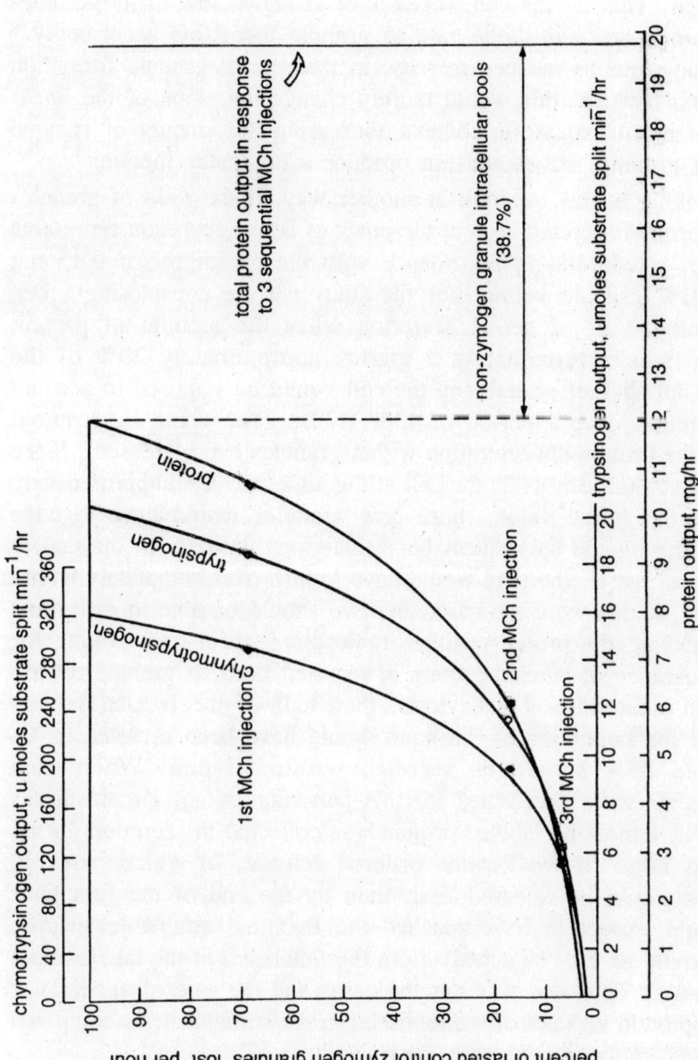

Figure 8.4 : Relationship between degranulation and chymotrypsinogen, trypsinogen and protein outputs. A breakdown of intracellular source is shown for protein secretion; 100% of granule contents of fasted rat pancreas only accounts for about 60% of total protein secretion.

linked to the number of granules present in the cell at a given time, the rate increasing as the number of granules per unit volume decreased. That is, the cell would have to know how many granules it contained and adjust the rate of granule formation accordingly.* A simple stimulus-induced increase in the rate of granule formation would not suffice; this would merely change the slope of the linear function in an exocytosis context, increasing the amount of enzyme lost per granule, but would not produce a nonlinear function.

Looking at this question in another way, if the ratio of granules lost to protein secreted seen at the onset of active secretion represents the true, fixed ratio in accordance with the paradigm (or 0.15 mg protein/1% granule volume for the study we are considering), then after some 3 hr of active secretion when the amount of protein secreted was approximately 3 mg/hr, approximately 20% of the original number of granules in the cell would be required to account for secretion over a period of 1 hr (N.B., less would be required only if the protein concentration within granules had increased). Since the total volume density in the cell at this time was less, approximately 17% of the initial value, more new granules would have to have formed during the subsequent hour than were initially in the cell at the outset, that is, the pool would have to turn over completely within the hour. If this were the case, then we should be able to determine it by labeling the proteins with a radioactive tracer and looking for the appearance of labeled protein in secreted fluid. If granule release occurred randomly by exocytosis, then half of the labeled protein made at the beginning of the hour should have been secreted by its end, and 25% should be secreted within 30 min. When such experiments were performed for this particular set of circumstances in the rat, almost no labeled protein was collected in secretion during the first hour. If we assume ordered release, in which case old granules would be released first, then by the end of the first hour we would expect to have reached the maximal attainable specific radioactivity as we "switched" from the unlabeled to the labeled pool of granules. This also was not the case, and the rate of appearance of new protein was actually diminished in the stimulated, as compared to the unstimulated state, not the opposite.

Furthermore, over the total period of study there was a decrease of some 95% in the volume density of zymogen granules, or in terms of the paradigm, in the number of granules. Therefore, whatever changes may have occurred, the rate of granule formation and loss were nevertheless very different, formation having occurred at only 5% of the rate of loss. That is, we can consider changes that occur in the granule pool of the cell over a period of a few hours primarily to reflect changes in the population of granules that were in the cell at the outset. In the context of exocytosis, this low rate of granule formation would hardly affect the observed relationship between the amount of protein secreted and the number of granules lost. Therefore, the *nonlinear relationship* between granules lost and protein secreted seen during the course of this study appears to reflect the kinetic pattern of secretion itself, not a balance between the formation of granules and their "*release*," and as such is not consistent with the transfer of material by a two-compartment exocytosistype system.

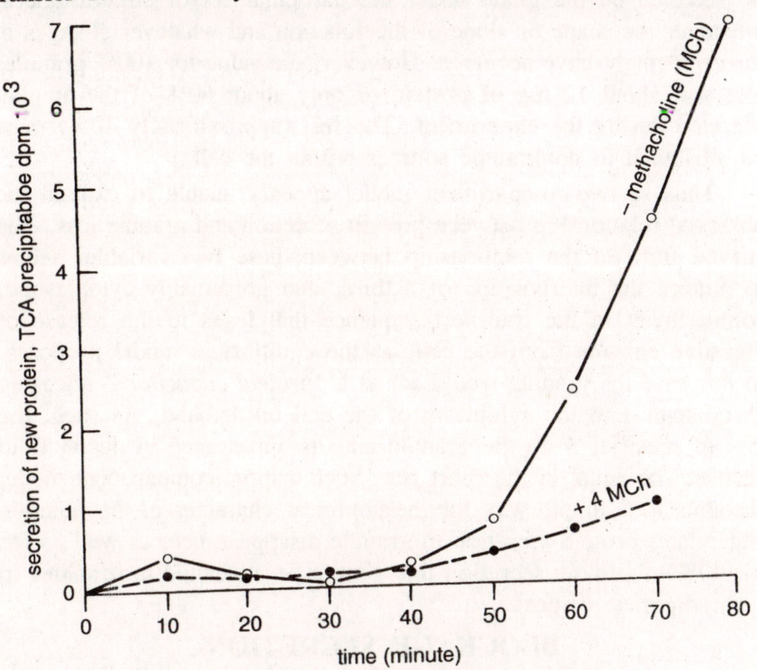

Figure 8.5 : Appearance of labeled protein in secretion after intravenous injection of labeled amino acids.

This view is reinforced by the following. Tissue content of amylase decreases after this course of stimulation by between one-third and one-half of that initially present. If we assume that this decrease applies to all secretory protein, then from an initial pool of some 70 mg/g tissue, we would lose between 23-30 mg/g of tissue, or more or less the same amount that was secreted (approximately 20 mg). That is, we have every reason to think that the material secreted was predominantly derived from a preexisting pool that was being depleted as a result of secretion, which, if it was being replenished, was being replenished at a relatively insubstantial rate compared with rate of secretion.

In addition to the question of the nonlinear nature of the function, if protein secretion were wholly accounted for by the direct transfer of material between two compartments, as in exocytosis, then more protein could not be secreted than could be accounted for by the loss of all the granules. In this case, 7 to 100% granule loss, then this value should correspond to the maximum amount of protein that could be secreted by the gland under the particular set of circumstances, whatever the shape or slope of the function and whatever changes in turnover might have occurred. However, the value for 100% granules lost was about 12 mg of protein or only about 60% of the amount secreted during the experiment. The rest (approximately 40%) must be attributed to nongranule sources within the cell.

Thus, a two-compartment model appears unable to explain the observed relationship between protein secretion and granule loss. The curved form of the relationship between these two variables seems to require the interposition of a third, and presumably cytoplasmic, compartment in the transport sequence that leads to the release of digestive enzyme from the cell, as the equilibrium model proposes. In this case the granule would act as a "*protein capacitor*," releasing its contents into the cytoplasm of the cell on demand. As such, the loss of material from the granule and its appearance in ductal fluid need not be equal in the short run. Such a three-compartment model accounts in a simple way for the nonlinear character of the function that relates protein secretion to granule disappearance, as well as for the 40% of protein secretion that cannot be attributed to granules in a paradigmatic context.

BLOCKAGE SECRETION

It was first noted many years ago by Grossman and his colleagues that the secretion of digestive enzyme by acinar cells was maintained at highly augmented rates, even though by light microscopy standards, granules no longer seemed to be present within them. Somewhat more recently, Jamicson and Palade also reported digestive enzyme secretion in the absence of zymogen granules. In the fasted rat after 3 hr of highly augmented protein secretion, the volume density of zymogen granules was decreased by about 95%. Therefore, these cells were essentially devoid of granules at this time (at least as estimated under the light microscope). However, digestive enzyme secretion nevertheless continued at a highly augmented rate (approximately one-third of the maximum or peak rate and a 10-fold increase over time-paired unstimulated controls). Thus, digestive enzyme secretion does not cease when the cell is emptied of its zymogen granules. Secretion can continue and at a highly augmented, although somewhat diminished, rate. Once again the prediction of the paradigm is not borne out by the evidence and a cytoplasmic origin for at least some secreted material seems indicated.

Table 8.2 : Enzyme Output from Degranulated Cells in Response to Methacholine Chloride.

		Enzyme Output (U/h)	
	Protein (mg/h)	Trypsinogen (μmoles substrate split min^{-1}/hr)	Chymotrypsinogen (μmoles substrate split -min^{-1}/hr)
Controls	0.23 ± 0.06	0.89 ± 0.18	3.45 ± 0.67
Methacholine	2.83 ± 0.32	6.51 ± 1.04	89.20 ± 13.75

Supporters of the paradigm have attempted to explain the secretion of enzyme in the apparent absence of zymogen granules by proposing that under these circumstances protein is secreted in much smaller granules (0.1 μm or less in diameter) that cannot be seen under the light microscope. This argument is based on the fact that small granules are seen in electron micrographs of depleted cells. In accordance with the paradigm they are viewed as new, not old smaller, granules, because granules of this size were not present initially. In any event, the number of such granules required to account for transport by exocytosis is substantially greater than for the larger zymogen granule. For example, the volume of a 1-μm diameter zymogen granule is about 0.5,43, whereas that of a 0.1-μm diameter granulc is 1000fold less. Therefore, 1000 times the number of granules would be required to move an equal amount of material (at the same concentration), and there is no indication that there are a sufficient number of small granules in "*depleted*" cells to accomplish such a task. In their absence, the rate of turnover of small granules would have to be prodigious in order to account for secretion. Of course, if the same number of small, as normal-sized, granules were present in the cell, then for a given amount of secretion, they would have to be formed and release their contents at 1000 times the rate. In addition, the amount of membrane that would have to be inserted into the plasma membrane would also be increased because the surface/volume ratio of spheres increases as diameter decreases. In this case (from 1 to 0.1 μm diameter), the surface-to-volume ratio would increase by 10-fold; that is, 10 times as much membrane would be required to move an equivalent amount of material. Thus, if we propose the exocytosis of smaller granules, then we must account for the fact that such transport would be much less efficient and would require magnitudinally greater numbers of granules, fusion events, and membrane to move an equivalent amount of material. Of course, if we propose even smaller granules, as has been suggested recently (the exocytosis of the small smooth-surfaced vesicle), the geometrical problems become even more severe and border on the ridiculous.

9

Coenzymes

A *coenzyme* is an organic molecule that can bind within the enzymes that require its function to *catalyze* a biochemical reaction. Hence, *coenzymes* are *organic cofactors* that augment the diversity of reactions that otherwise would be more limited to chemical properties of side chain substituents from amino acid residues within protein enzymes. Coenzymes bind to apoenzymes to generate *catalytically* competent *holoenzymes*. With tight binding, coenzymes may be referred to as *prosthetic* groups; with loose binding, they may be called cosubstrates.

GENERAL HISTORY

Most work on the chemical nature of coenzymes began in the 1930s, when the structures of some vitamins in the B complex were being elucidated. In 1932 Auhagen had succeeded in dissociating a heat-stable component of the yeast "*carboxylase*," which he called "*cocarboxylase*." By 1937, Lohmann and Schuster had determined its structure as *thiamine pyrophosphate* (TPP). During the same period Warburg and Christian were active in elucidating the nature of the oxidation-reduction coenzymes that contain *riboflavin* (FMN, FAD) and nicotinamide (NAD, NADP). By the mid-1940s, the coenzyme forms of vitamin B_6 (PLP, PMP) were recognised by Gunsalus, Snell, and others. The discovery by Lipmann in 1947 of a coenzyme of acetylation (CoA), which contained pantothenate, led to full structure elucidation by Baddiley et al. in 1953. Advances also made in the 1950s by a number of laboratories led to recognition of coenzyme forms of folacin, and by the end of the decade, Lynen et al. had characterised covalent 1'-N-carboxybiotin and Barker and his associates had identified a coenzyme form of vitamin B_{12}. Prerequisite to determining structures of coenzymes was recognition and purification of enzyme systems that depend on the catalytic participation of these organic cofactors. Continued examinations of diverse enzymes are extending the list of newer coenzymes, some of which do not derive from vitamins.

INDIVIDUAL COENZYMES

Vitamin-Derived Coenzymes

Many coenzymes are the metabolic result of converting ingested vitamins, especially those of the B complex, to forms suitable for binding and function in enzyme systems. Of the 13 vitamins presently known to be required in the diet of humans and most animals, at least eight (thiamine, riboflavin, niacin, vitamin B_6, folacin, vitamin B_{12}, biotin, and pantothenate) are simply the essential precursors for coenzymic forms made in the body.

Other Coenzymes

A growing number of coenzymes are not formed from vitamins; hence they can be biosynthesized by at least most of the organisms that require them. Several additional coenzyme types are described briefly.

Pyruvoyl functions

A simpler, but less frequently encountered, variation on one way in which pyridoxal 5'-phosphate operates is provided by the pyruvoyl N-terminus of some enzymes. In these *electrophilic* centers, the amino function of a substrate amino acid condenses to form a ketimine, which facilitates loss of a carboxyl group from the attached amino acid moiety. For example, S-adenosylmethionine decarboxylases from mammals as well as *Escherichia coli* use such a system to form *spermidine* from *putresine* and *methionine*.

Lipoyl functions

α-Lipoic (thioctic) acid occurs naturally in amide linkage to the e-amino group of lysyl residues within transacylases that are core protein subunits of α-keto acid dehydrogenase complexes. In such cases, the functional dithiolane ring is on an extended flexible arm. The lipoyl function mediates the transfer of the acyl group from α-hydroxyalkyl-TPP to CoA in a cyclic system that transiently generates the dihydrolipoyl residue.

Ubiquinones

The ubiquinones (coenzyme Q's) are a group of "ubiquitous"

Table 9.1 : Coenzymatic Forms and Functions of B Vitamins

Coenzyme Forms	*Vitamin Enzyme Systems*	*Functions*
	Thiamine (B_1)	
Thiamine pyrophosphate (perhaps triphosphate)	α-Keto acid decarboxylases, transketolase	Decarboxylations of α-keto acids from metabolism of carbohydrates and amino acids; glycolaldehyde transfers; nerve membrane ion transport
	Riboflavin (B_2)	
Flavin mononucleotide, flavin-adenine dinucleotide (covalent and noncovalent)	Oxidases, dehydrogenases	Hydrogen and electron transfers from substrates to oxygen and to cytochromes.
	Niacin	
Nicotinamide adenine dinucleotide and the 2' phosphate	Reductase, dehydrogenases	Hydride ion transfer
	Vitamin B_6	
Pyridoxal 5'-phosphate, pyridoxamine 5'-phosphate	Aminotransferases, amino acid decarboxylases, amino acid dehydratases, cystathionine synthetase, phosphorylase, 5-aminolevulinate synthetase, etc.	Transaminations, decarboxylations, and side-chain cleavages of amino acids; breakdown of glycogen; biosynthesis of heme, phospholipid, etc.
	Folacin	
Tetrahydrofolate and γ-glutamyl conjugates	Transformylases, thymidylate synthase hydroxymethyl-	One-carbon activation and transfer as in purine and pyrimidine biosynthesis,

	transferases, formiminotransferase, etc.	catabolism of serine and histidine, etc.
	Vitamin B_{12}	
5'-Deoxyadenos--ylcobalamin	i-Methylmalonyl-CoA mutase	Isomerization of i methylmalonyl-CoA to succinyl-CoA
Methylcobalamin	5-Methyltetrahydrofolate-homocysteine methyltransferase	Methyl group transfer as in methionine biosynthesis
	Biotin	
I'-N-Carboxybiotin as amide-linked to ε-lysyl residues	Acyl-CoA carboxylases and other bicarbonate -dependent carboxylases	Co_2 activation and transfer in formation of acids
	Pantothenic Acid	
CoA	Acyltransferases, etc.	Fatty acid metabolism, etc.
4'-Phosphopantetheine	Acyl carrier protein of fatty acid	Fatty acid biosynthesis synthetase

substituted benzoquinones with variable-length terpenoid chains. In eukaryotes, they are found mainly in the *mitochondrial* inner membrane, where they function to accept electrons from several different dehydrogenases and relay them to the *cytochrome system*. In this process, the quinone function can be cyclically reduced and reoxidised. Recently evidence has been found for a function of CoQ in plasma membrane electron transport, which influences *mammalian* cell growth.

Pyrroloquinoline quinone and quinoproteins

The coenzyme *pyrroloquinoline quinone* (PQQ) originally was isolated from bacteria possessing a *methanol dehydrogenase*. PQQ functions as a redox component of holoenzymes in which the coenzyme can be reduced in some cases by successive one-electron steps to the radical and dihydro forms and in other cases directly to the dihydro level by two-electron processes. More recent detailed examinations of several quinoproteins have elucidated other quinone-hydroquinone-related coenzymes as side chain altered prosthetic groups. Examples include the 6 hydroxydopa (trihydroxyphenylalanyl),residue at the active site of serum amine oxidase and the *tryptophan tryptophylquinone* in a bacterial methylamine dehydrogenase.

Methanogen coenzymes

There are a diverse group of coenzymes required to convert carbon dioxide to methane-rather novel compounds that vary from a complex nickel-porphyrin-like F430 to the phenyl ether of methanofuran, the unusual pterin of *tetrahydromethanopterin*, the deazaflavin of F_{420}, to the somewhat simpler sulfur-containing *mercaptoheptanoylthreonine* and coenzyme M. These coenzymes are used by highly specialised *archaebacteria* (known as methanogens) that can accomplish the reduction of CO_2 through *formyl*, *methyl*, *methylene*, and *methyl* stages of CH_4. The importance of the coenzymes that participate can be appreciated by the quantitative and global aspects of carbon cycling through methane-forming systems, which are not only free-living microbes but also are found in the human intestinal tract.

10

ISOENZYMES

Isoenzymes (Greek: *izos,* equal; *zymos,* leaven), also known as isozymes, are enzymes that catalyze the same reaction but are different in structure and often in some of their properties. Isoenzymes are proteins encoded by distinct genes that can be at different loci or can represent different *alleles* at the same locus. The Commission on Biological Nomenclature of the International Union of Biochemistry recommended that the term "*isoenzyme*" be restricted to forms of the same enzyme that originate at the level of different genes. The discovery of the heterogeneity of human lactate dehydrogenase (LD) in 1957, as well as the introduction of *histochemical* techniques for the detection of esterases, led to the proposal of the term "isozyme" by Markert and Moller in 1959 to describe different proteins having similar enzymatic activity.

Differences between various forms of the same enzyme can be genetically determined or can be due to epigenetic changes of the enzyme. Although the latter forms are not isoenzymes, from an operational point of view they have been often called isoenzymes. Newly discovered variants of enzymes cannot be considered isoenzymes until their genetic control is determined. However, this term is sometimes used in a general sense to signify different forms of an enzyme regardless of the mechanisms generating the enzyme diversity.

DERIVATION OF ISOENZYMES

Many enzymes are coded for by multilocus genes. The genes that code for specific isoenzymes can be on the same chromosome (e.g., chromosome 4 for the four loci of alcohol *dehydrogenase* in humans) or on different chromosomes (e.g., the hexosaminidase alpha and beta subunits on human chromosomes 15 and 5, respectively). Isoenzymes can also originate from mutations of alternative forms of genes (alleles) at the same locus ("*allelozymes*": e.g., allelic variants of *placental alkaline phosphatase*), or they can arise from the modification of the structure or expression of genes in somatic cells-this can result from *malignant* transformation.

Isoenzymes made up of subunits that are associated in various combinations, producing hybrid molecules when the subunits are different, constitute *oligomeric isoenzymes*. These molecules may be generated by random associations of units derived from different loci or from multiple alleles at the same locus [e.g., creatine kinase (CK), lactate dehydrogenase (LD), and hexosaminidase]. *Lactate dehydrogenase* in serum and organs of vertebrates is a tetramer of two different *polypeptide* subunits, A or H (from heart) and B or M (from muscle), which combine randomly, producing *homopolymeric* and *heteropolymeric* molecules. Thus, five different LD isoenzymes are formed, from he most anodal LD-1 (four H subunits), to the most cathodal LD-5 (four M subunits). Lactate dehydrogenase-2, LD-3, and LD-4 are hybrid molecules, with subunit composition H_3M, H_2M_2, and HM_3, respectively. An additional LD isoenzyme, LD-C, composed of four identical subunits (C), which are different from the other two subunits of LD, is characteristic for postpuberal testis (*primary spermatocytes*). In man, *hexosaminidase* exists as two main forms: B, with a structure of four beta subunits, and A, with a structure of one alpha and two beta subunits. Because of their different genetic control, the isoenzymes differ in their amino acid structure; however, the amino acid sequence of the active site is usually highly conserved and is the same for different isoenzymes of a single enzyme.

Isoenzymes that are produced as a result of allelic variations (*allelozymes*) are inherited codominantly; that is, the products of both alleles are expressed. The patterns of these isoenzymes can thus be used to trace their genetic inheritance. On the other hand, isoenzymes that have arisen from genes at multiple loci have been distributed throughout the species population over the course of evolution, resulting in all individuals possessing essentially the same complement of isoenzymes. The existence of multiple gene loci for specific isoenzymes can be explained by two different evolutionary processes. In one process, catalytic activity may have arisen from two or more unrelated events leading to the establishment of two different genes, each with the capacity to code for a different enzyme protein that catalyzes the

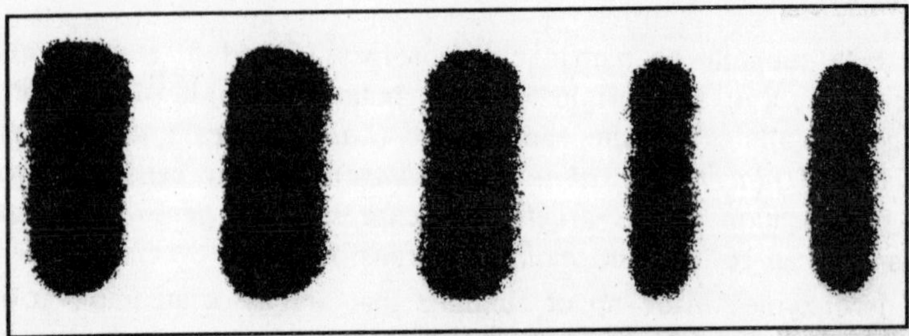

Figure 10.1 ::actate dehydrogenase (LD) isoenzymes from normal human serum. The anode is at the left and the most anodic band is LD-1. The proteins were separated by zone electrophoresis in a thin layer of agarose gel in 0.01 mol barbital buffer, pH 8.6.

same reaction. The two genes thus formed can undergo mutations and will then code for different enzyme-proteins. In the other, most common process, a single ancestral gene coding for a single enzyme is duplicated. The two resultant genes can mutate and would then be modified favorably for the organism, cell, or organ function that was evolving. This latter process is responsible for the occurrence of isoenzymes controlled by genes at different loci. Although the occurrence of major change in the gene products over time could be predicted, the similarities in the two gene products should be great enough to permit one to ascertain their common ancestry. Such ancestral interspecies homologies exist, for example, in the isoenzymes of LD. For some isoenzymes (e.g., human a-amylase), the two loci occurring after gene duplication remain closely linked.

Genetically determined changes in isoenzymes between individuals can explain abnormalities in metabolism manifested as hereditary metabolic diseases, and can also account for individual sensitivity to various drugs.

STRUCTURE, FUNCTION, AND DEVELOPMENTAL ASPECTS

By definition, isoenzymes must differ in primary structure to some degree. Full elucidation of these differences requires the application of physical and chemical methods of protein analysis to purified isoenzyme preparations. The complete amino acid sequencing of isoenzymes has been achieved for various isoenzymes (e.g., human aldolases A and B; human carbonic anhydrases I and II, which differ in about 90 of 260 amino acid residues). A sensitive and specific method for analyzing subtle differences in primary structure among isoenzymes is the "*fingerprinting*" technique, which compares two-dimensional maps of the peptides in a partial hydrolysis of the enzyme-protein. This method is especially useful for identifying single amino acid substitutions in products of allelic genes. The elucidation of the secondary and tertiary structures of isoenzymes has been achieved by X-ray crystallography.

Since isoenzymes are coded for by different genes, they are expected to have different properties such as substrate concentration optimum, *electrophoretic mobility*, resistance to inactivation, sensitivity to inhibitors, Michaelis constant for substrate(s), relative rate df activity with substrate analogues, and sometimes a different extent of substrate specificity. Isoenzymes of LD and *carbonic anhydrase*, for example, can have different functions because of different properties. The difference in charge distribution over the surface of the molecule can affect the location of the molecule within the cell.

Some isoenzymes show tissue or cell specificity, and individual isoenzymes can be predominantly represented in cells at each stage of their differentiation, from embryo to adult. For example, changes in LD and CK isoenzyme distribution occur during development. The relative proportions of cytoplasmic and mitochondrial forms of certain isoenzymes also change during differentiation. Isoenzymes can differ in their substrate specificity. When substrate specificity is not stringent, isoenzymes can show differences in their activity with substrate analogues; for example, alkaline phosphatases of bone, liver, and kidney differ in this respect from those of placenta and small intestine. Altered substrate specificity has also been shown for isoenzymes of mammalian liver alcohol dehydrogenases and for acid phosphatases (e.g., in rabbit muscle).

The presence of different isoenzymes in certain tissues, cells (e.g., enolase, LD, *phosphofructokinase*, CK, *hexokinase*, *pyruvate kinase*, *amylase*) or subcellular structures such as lysosomes and mitochondria (e.g., the NAD-dependent malate dehydrogenase, isocitrate dehydrogenase, and adenylate kinase isoenzyme systems, with one isoenzyme in the mitochondrion and another in the cytosol), implies a specialised role for each isoenzyme in cell metabolism. Indeed, isoenzymes allow a fine adaptation of metabolic patterns to ontogenic development or to changes in the environment. One physiological advantage of the isoenzymes is the creation of a compartmentalisation of the same catalytic reaction in different metabolic pathways. Compartmentalisation allows precision in the maintenance of metabolic patterns unachievable by a single enzyme. Isoenzymes of LD, for example, are represented in different percentages in various organs; this differential distribution has functional significance. A correlation exists between the properties of isoenzymes predominant in certain tissues and the metabolic patterns in these tissues. There are also age-related changes in isoenzymes within certain cells. The proportions of the isoenzymes in a tissue or organ depends on their half-lives (i.e., rates of synthesis and degradation).

Many isoenzymes are named according to their tissue or subcellular structure of origin (e.g., placental and liver alkaline phosphatase); isoenzymes that can be separated by electrophoresis are numbered according to their electrophoretic mobility. Because most enzymes are negatively charged, the most anodic isoenzyme is numbered as the first.

METHODS OF ANALYSIS

Isoenzymes can be separated by various techniques-zone *electrophoresis* as well as capillary electrophoresis, elution-convection electrophoresis, isoelectric focusing, or discontinuous *electrofocusing*. These methods and various types of ion exchange chromatography take advantage mainly of differences in the net electric charge of isoenzyme molecules at a given pH and, to some extent, of differences in molecular size and shape. However, the separation of certain isoenzymes requires other methods such as high performance liquid *chromatography* (HPLC) or affinity chromatography. Once separated, the isoenzymes can be identified on the basis of some of their properties. For *colourimetric* detection in serum or tissue extracts, histochemical techniques are used to show *enzymatic* activity by producing a coloured precipitate or a fluorescent substance when fluorogenic substrates are used or when the product of the enzymatic reaction is fluorescent. This technique, known as the *zymogram* technique, or *enzymoelectrophoresis*, is commonly performed in clinical laboratories for diagnostic purposes, with the use of scanning densitometers and fluorimeters. Other methods make use of inhibition or denaturation of enzymes by heat, concentrated urea, or other organic compounds; reactivity with different substrates, coenzymes analogues; or measurements of pH optima, Michaelis constants, and so on.

Immunological techniques are also used to separate and to quantitate isoenzymes because most of them differ with respect to antigenicity. Monoclonal antibodies raised against pure isoenzymes are particularly useful to study isoenzyme subunit structure and can detect subtle differences between various forms of enzymes. Immunoinhibition techniques (i.e., measurement of the residual activity after treatment with specific antibodies) can be used for isoenzyme quantitation. Immunological precipitation followed by measurement of the enzyme catalytic activity or of the amount of radioactive label bound to the enzyme allows quantitation and expression of isoenzyme amount either in units of activity or in mass units.

Recently, genetic (protein) engineering techniques, including oligonucleotide-directed, site-specific mutagenesis and chimeric fusion protein formation, have been used to study structure and function of isoenzymes (e.g., human aldolases A and B). With the advent of the transgenic technology, new isoenzyme systems can be made-e.g., by inserting the required genes into the genome, even at specific locations in the genome (by homologous recombination techniques). The genes for the new isoenzymes can be altered to elicit function in specific cells at certain times in development.

PRACTICAL APPLICATIONS OF ISOENZYME ANALYSIS

lsoenzymes have not only theoretical importance for understanding basic biological processes (e.g., the transformation of one gene into another, the regulation of gene expression, and the significance of metabolic pathways in different tissues), but also their analysis has a broad array of practical applications. These wide-ranging applications include use as markers in human-mouse cell hybrids made for human chromosome mapping and as diagnostic indicators in both hereditary and nonhereditary diseases. Because isoenzymes are valuable markers of gene function, they can be used to study population genetics (e.g., to identify new phenotypes and genetic polymorphisms). With isoenzyme analysis, the measurement of allele frequencies in a population facilitates the study of evolutionary selective pressure.

lsoenzyme analysis has also been used in forensic science, to distinguish species and tissues, and in paternity testing, based on allelic variations as an index for testing.

lsoenzymes can be used for studying cell differentiation and alterations of this process. Because new patterns of isoenzymes commonly occur when fully differentiated cells undergo malignant transformation, isoenzymes are valuable for the diagnosis of malignancy as well as for understanding some mechanisms of malignant transformation. Examples of some ubiquitous isoenzymes that are altered in tumors include aldolase, LD, and (3-hexosaminidase.

The organ or tissue specificity of isoenzymes allows them to be used to pinpoint organ damage in various diseases. For example,

CK-2 (CK-MB) isoenzyme is increased in the serum of patients with myocardial damage. The diagnostic specificity is further increased by simultaneously measuring several different isoenzymes in serum and other body fluids.

A significant number of relatively rare inborn errors of metabolism result from mutations leading to the deficiency or modification or a specific isoenzyme. The functional consequences of these mutations vary depending on the nature of the alteration. When allelic variation of isoenzymes results in disease, usually the catalytic activity of the mutant enzyme is greatly reduced or absent. The recognition of a specific isoenzyme deficiency not only provides a firm biochemical basis for a clinical diagnosis but also serves as a means to distinguish variants of some metabolic disorders. Ultimately, the biochemical classification of subtypes leads to an understanding of the molecular pathology of individual forms of a disorder.

As a result of mutations at multiple gene loci, a number of disorders arise from alterations of isoenzymes coded for by these loci. The metabolic effects of such alterations consequently are manifested only in the tissues or organs in which the affected isoenzyme is present. For example, in muscle disorders there is a tissue-specific isoenzyme deficiency in either glycogen or fatty acid metabolism leading to symptoms of skeletal muscle injury.

Some lysosomal isoenzyme deficiencies can lead to progressive neurodegenerative disease usually (but not always) manifested in childhood. The best known example is Tay-Sachs disease (deficiency of the alpha subunit of (3 hexosaminidase A). Examples of other lysosomal storage disorders in which a single genetically distinct isoenzyme is affected include mannosidosis (a-mannosidase A deficiency), metachromatic leukodystrophy (arylsulfatase A deficiency), and Maroteaux-Lamy disease (arylsulfatase B deficiency).

NONISOENZYMIC MOLECULAR FORMS OF ENZYMES

Posttranslational modification of the enzyme-protein can also result in multiple molecular forms of enzymes. Although these forms are not isoenzymes (i.e., they are not coded for by separate genes), historically and from an operational point of view they have often been referred to as isoenzymes because they have many characteristics similar to isoenzymes-that is, they can exist in different cell organelles and in different metabolic compartments within a single cell (e.g., the cytoplasmic and mitochondrial forms of aspartate aminotransferase).

Posttranslational mechanisms of producing enzyme variants include the formation of aggregates of a single unit (homopolymers) or of an enzyme with nonenzymatic proteins (e.g., cholinesterases), and other epigenetic modifications of the initial protein such as combination of the protein with nonprotein molecules (e.g., sialic acid residues). Enzymes with increased mass and altered electrical charge arise when they bind to immunoglobulins ("*macro*" enzymes: e.g., *macroamylase* and *macro-CK*). Alteration of carbohydrate side chain, acylation, deamidation, sulfhydryl oxidation, and partial cleavage of the polypeptide chain are other mechanisms of producing *nonisoenzymic* molecular forms of enzymes. The removal of some residues can explain differences in function because, for example, deamidation and phosphorylation may influence secretion and uptake of enzymes by cells. Partial proteolysis of the initial enzyme polypeptide can produce so-called isoforms of enzymes when only one amino acid is removed [e.g., isoforms of CK-2 (CK-MB), CK-MB,, and CK-MB,, which differ by a ter ninal lysine and can be separated electrophoretically]. It is also conceivable that conformational changes of the polypeptide chains can lead to "*conformational enzymes*" or "*conformers*," with the same molecular weight and amino acid sequences but different properties, such as changed electrophoretic mobility. The different forms of enzymes that are generated through posttranslational modifications of the enzyme-proteins in general do not differ in their catalytic characteristics but have biological significance and practical applications similar to those of isoenzymes. Once their genetic control has been elucidated, some of the different molecular forms of enzymes may prove to be, in fact, isoenzymes.

11

CYTOCHROME

Cytochrome P450 is the generic name applied to a large superfamily of *hemoprotein*, mixed-function oxidases that *metabolise* a structurally diverse group of *exogenous* and *endogenous* organic substrates. The name is derived from the prominent absorption band observed at 450 nm following reduction of the heme iron and its coordination with carbon monoxide. These enzymes are widely distributed among *microorganisms*, plants, insects, fishes, and animals, and they catalyze a reaction of the general nature:

$$\underset{}{O_2} + \underset{\text{biological reducing equivalents}}{NAD(P)H} + \underset{\text{organic substrate}}{AH} \xrightarrow{P450} \underset{\text{hydroxylated organic product}}{AOH} + H_2O + NAD(P)$$

In the case of *exogenous* substrates, this reaction most frequently is involved in an organism's effort to detoxify foreign compounds (xenobiotics) derived from the environment. In the case of *endogenous* substrates, the reaction is generally involved in the production of biologically more active compounds (e.g., steroid hormones) from less active compounds (e.g., *cholesterol*).

CHARACTERISTICS OF P450s

Ryo Sato and Tsuneo Omura described in 1962 an unusual pigment in a subcellular fraction of rabbit liver, the *endoplasmic* reticulum, which upon reduction and coordination with carbon monoxide showed an intense absorbance band at 450 nm. Hence the name, pigment 450 nm or P450. This substance was subsequently clearly established to be a protoheme-containing protein, the same *prosthetic* group associated with *hemoglobin*. However, P450 and hemoglobin are quite distinct in that when reduced and coordinated with CO_2 *hemoglobin* has an intense absorbance band at 420 nm. It is now apparent that P450 has a thiolate ligand provided by cysteine in the fifth coordination position of the heme iron, while hemoglobin contains an imidazole ligand from histidine at this coordination position. This difference is the basis of the spectral difference between these two heme proteins. Functionally, hemoglobin serves to reversibly bind and transport oxygen and therefore does not reduce O_2, while P450 function hinges on the ability to reduce oxygen during the mixed-function oxidation reaction. The thiolate ligand participates by lowering the redox potential of the heme iron in P450, thereby facilitating oxygen reduction.

In bacteria, cytochromes P450 are soluble in the cytoplasm, while in all higher species including plants, P450s are firmly anchored in membranes. Most forms of P450 are found in the endoplasmic reticulum, where they face the cytosol. A few forms of cytochrome P450 are localised to the inner *mitochondrial membrane*, where they face the matrix.

Cytochromes P450 usually cannot function on their own. They require transfer of reducing equivalents from reduced pyridine *nucleotides* (NADH or NADPH) to the P450 heme iron. In bacteria, virtually all forms utilise electrons from NADH, which are transferred via an electron transport chain consisting of a *flavoprotein* that extracts electrons from NADH and passes them on to an iron-sulfur protein, which in turn reduces the P450. Mitochondrial P450s represent a variation on this theme whereby electrons are transferred from NADPH via a flavoprotein and iron-sulfur protein located in the mitochondrial matrix. P450s in endoplasmic reticulum (microsomes) derive their reducing equivalents from NADPH via a *ubiquitous flavoprotein* known as NADPH cytochrome P450 reductase. This flavoprotein is also firmly embedded in the membrane of the endoplasmic reticulum. In animals, there appears to be a single form of P450 reductase, which services a rather large number of different forms of microsomal P450.

Thus one way of classifying P450s is based on the manner in which electrons are transferred.

Bacterial and Mitochondrial P450s:

NAD(P)H → flavoprotein → iron-sulfur protein → P450
(FAD-containing) (2Fe-2S protein)

Microsomal P450s:

NADPH → flavoprotein → P450
(FAD + FMN-containing)

Recently an interesting variation of this *microsomal* P450 system has been characterised as a soluble protein in bacteria. A P450 protein is extended at its carboxyl terminus by a domain having most of the characteristics of the microsomal P456 reductase. This fusion protein is the sole known bacterial P450 to resemble the microsomal P450 system rather than the mitochondrial system and is the single form of P450 able to function on its own with only the presence of NADPH.

ACTIVITIES OF P450s

Virtually every animal cell contains one or more forms of P450. Another way of classifying P450s has been based on their enzymatic properties. In the broadest sense, P450s can be grouped into forms that metabolize endogenous substrates and forms that metabolize exogenous substrates.

The forms involved in endogenous substrate metabolism tend to be localised to specific functional sites such as steroidogenic tissues, (adrenal, gonads, placenta) or a specific organ such as the kidney or liver, while the forms involved in exogenous substrate metabolism are sometimes widely distributed. Sites of exogenous metabolism are found in every organ including the skin, with highest concentration and numbers of these forms of P450 being found in the liver. Recently is has been learned that unique P450s can exist in only a limited number of cells such as the nasal epithelium, and it is imagined that many forms of P450 yet undiscovered reside in such limited locations.

P450-mediated *endogenous* substrate *metabolism* includes steps in *cholesterol biosynthesis* (14-demethylase), biosynthesis of steroid hormones from cholesterol (*progesterone*, *aldosterone*, *cortisol*, *testosterone*, *estrogen*), bile acid biosynthesis, fatty acid hydroxylation, and arachidonic acid metabolism. These reactions lead to production of key regulators of biological processes such as reproduction and vascular activity. Other potential roles for P450s include biosynthesis of neurosteroids and activation or inactivation of important compounds in growth and development, including derivatives of vitamin A. In the best studied of the endogenous systems, that involved in biosynthesis of steroid hormones, it is clearly established that amino acid mutations within the P450 polypeptide chain lead to genetic diseases that alter the homeostasis and even the phenotype of the individual. It can be presumed that genetic diseases will be found to be associated with most or all of the forms of P450 metabolizing endogenous compounds.

While the endogenous activities just listed clearly define the biological significance of the P450 systems, P450-mediated metabolism of xenobiotic compounds is equally important. This too, represents a broad scope of activities ranging from use of environmental compounds as the sole carbon source in bacteria to the biotransformation and detoxification of xenobiotics in animals. As a general consideration, these forms of P450 may metabolize structurally diverse chemicals and demonstrate lower substrate specificity than do the forms involved in metabolism of endogenous compounds. Types of biotransformations include N-, S-, and Chydroxylations and dehalogenations, as well as deaminations, dealkylations, and reductions of countless drugs, chemical carcinogens, mitogens, and other environmental contaminants. Included in xenobiotic biotransformation is drug metabolism, an issue of considerable interest to the pharmaceutical industry, particularly in light of the variability in the levels of different P450s in different people. In studies with laboratory animals, inbred strains are generally used in which all individuals contain relatively constant levels of the different forms of P450. Humans are, of course, outbred, and each individual can be considered to have his or her own unique P450 profile. Consequently the effectiveness of drug therapy may be dependent on thys profile, the individual P450 pattern being dependent on genetic (sex, age) as well as environmental (nutrition, exposure to foreign compounds) factors. In some cases different forms of P450 metabolize the same chemical substrate by different reactions, leading to different patterns of products. Thus individual variation in the levels of different P450s can significantly affect drug metabolism, particularly in instances of combined drug therapy. Clear examples of these individual variations are seen in the polymorphisms that exist within the human population with respect to P450-dependent metabolism of certain drugs, such that specific individuals will be defined as poor metabolizers (i.e., they do not clear the drug as efficiently as the general population, with resultant unwanted effects reminiscent of drug overdose).

While the detoxification of foreign compounds is generally beneficial to animals, in some cases reactive intermediates produced by P450 metabolism can be toxic, mutagenic, or carcinogenic. For example, benzo[a]pyrene, a polycyclic aromatic hydrocarbon in tobacco smoke, is metabolised by a specific form of P450 into a carcinogen that is thought to play a key role in lung cancer.

HOW MANY P450s ARE THERE?

The P450 superfamily is both large and evolutionarily ancient. While subcellular location and enzymatic activity are important characteristics of the P450s, these properties do not provide the most useful approach to classifying this large number of proteins. Rather, a systematic classification for P450s has been developed based on amino acid sequence. This scheme leads to classification into gene families and subfamilies based on primary sequence relatedness. Thus the 30 known human P450s are distributed among 12 gene families scattered among 9 different chromosomes. In addition to the mammalian P450s, unique gene families have been found in insects, snails, yeast, fungi, plants, and bacteria. The total number of known P450s in biology exceeds 240 and is rapidly increasing as new forms are identified.

In humans, the most abundant forms of P450 have already been identified. Nevertheless, we can expect that a substantial number of forms that are localised in specific cell types in the brain, intestine, and other organs remain to be discovered. While these may be of relatively low abundance compared to many of the forms already known, they will be surely found to play very important roles. For example, in the nasal epithelium unique forms of P450 have been found which presumably play important roles in *detoxification* and protection from *desensitisation* of the sensory system involved in smell. With the application of the powerful techniques of molecular biology to the identification of P450s, the sequences of proteins are easily determined by cloning, but the substrate specificity and therefore the function of these proteins cannot be so easily elucidated. Thus we now know the sequence of several P450s for which the function remains unknown. It may be that there will be as many as 200 different forms of P450 identified in humans. Even though P450s have related amino acid sequences and, in fact, contain diagnostic sequence motifs, universal P450 probes (either *immunological* or *oligonucleotides*) have not yet been developed. Thus we can expect that it will take several more years before the majority of the rare human P450s are identified.

Since eukaryotic forms of P450 are integral membrane proteins, they are difficult to release from their membrane environment and subsequently purify. Thus it has been difficult to obtain sufficient quantities of these P450s for biophysical investigation of their structure-function relationships. Accordingly, the only detailed structural information available on members of the P450 superfamily is from soluble, bacterial P450s. While this information provides a general picture of P450 structure, until the detailed structure of *eukaryotic* forms of P450 is determined we will not know for certain the common structural features associated with all P450s. Nevertheless, the ability to carry out site-directed mutagenesis based on predictions from bacterial P450 structural analysis, coupled with heterologous expression of cDNAs in systems having low or absent background levels of P450, is leading to detailed information on how P450s work.

REGULATION OF P450S

Regulation of gene expression is one of the key biological processes controlling development, tissue specificity, and homeostasis. The developmental roles of the P450 superfamily are not yet well defined. In a few instances it is clear that the timely expression of specific genes encoding P450s is essential; for example, formation of male secondary sex characteristics requires expression of P450s involved in *testosterone* synthesis at a specific moment in early fetal life. Certainly it will be found that various forms of P450 play important roles in different aspects of growth and development, including metabolism of *endogenous* compounds that serve as signals for different growth processes. Likewise, the molecular basis of tissue-specific expression of P450s is not yet well understood, but many forms are found only in certain cell types, indicating that complex regulatory processes are important at this level as well. The maintenance of P450 activities associated with endogenous substrate *metabolism* throughout adult life is known to be controlled by other endogenous compounds-for example, peptide hormones and steroid hormones. However, the detailed biochemistry of regulation of these genes is not yet understood.

One of the intriguing aspects of the P450 system is that many forms involved in the metabolism of exogenous compounds are induced (increased) by exogenous compounds. Often substrates for specific forms of P450 will induce these forms; for example, the forms of P450 that metabolise polycyclic *aromatic hydrocarbons* are induced by polycyclic aromatic hydrocarbons. Thus the environment regulates the drug-metabolizing profile in individuals. In some of these instances a particular P450 is undetectable until the individual is challenged by specific xenobiotics, while in other cases lower levels are increased to higher levels by such challenge. In summary, the regulation of the P450 profile in individuals will prove to be a very complex process involving developmental, tissue-specific, endogenous, and exogenous factors.

FUTURE DIRECTIONS

Many questions remain to be answered concerning the P450

superfamily in relation to the number of P450s, their structure-function relationships, their roles in key biological process such as growth and development, and the basis on which expression of P450 genes is regulated. In addition, the application of the unique P450-dependent chemistry to the production of fine chemicals or to the removal of environmental contaminants using microorganisms and engineered P450s in *heterologous* systems can be anticipated. Finally, we can foresee a time when P450 profiles will be analyzed in a noninvasive fashion and drug regimens for treatment of diseases will be tailor-made for the individual. The P450 superfamily will surely continue to be a major focus of biomedical research, yielding intriguing new discoveries well into the twenty-first century.

12

PHOSPHOLIPASES

Phospholipases comprise a subclass of lipases that hydrolyze phospholipids. There are many different phospholipases. They are defined by the position they attack on the phospholipid molecule. Phospholipases are central enzymes in both lipid metabolism and signal transduction. The products of phospholipase hydrolysis are involved in such diverse events as the mobilization of intracellular calcium (inositol trisphosphate), the regulation of protein phosphorylation (diacylglycerol), the disruption of membrane integrity (lysophospholipid), the activation and aggregation of platelets (platelet-activating factor), and the onset of inflammation (arachidonic acid metabolites: prostaglandins and leukotrienes). Many current research efforts are directed at gaining insight into the mechanisms of action of these proteins and the control of their activities.

PHOSPHOLIPIDS

Phospholipids have long been recognized as the major structural components of biological membranes. More recently, it has become apparent that phospholipids and phospholipid-derived mediators are involved in signal transduction and the control of cell activation and homeostasis.

As shown in Figure elsewhere in this chapter, the backbone of a phospholipid is a threecarbon molecule called glycerol. The stereospecific numbering positions sn-1 and sn-2 of the glycerol molecule are occupied by either long chain fatty acids or, in some instances, fatty alcohols. A polar head group is attached to the sn-3 position through a phosphate-oxygen bond. The head group can be phosphocholine, -serine, -ethanolamine, -inositol, -glycerol, or hydroxyl (phosphatidic acid).

PHOSPHOLIPASE SPECIFICITY

Phospholipases are part of a family of enzymes called hydrolases. Hydrolases use water to catalyze the degradation of biological molecules to their component parts. Hence, phospholipases use water to degrade phospholipid molecules. The reaction catalyzed by phospholipase A_2 is an example of this type of catalysis.

Phospholipase attack can occur at a carbon-oxygen or phosphate-oxygen bond. Phospholipases are defined according to the bond that is broken during catalysis. A great deal of information is available about phospholipases A_2, C, and D.

Less is known about phospholipase A,, which breaks the bond at the *sn-1* position of phospholipids. Many enzymes are known to have both triglyceride lipase activity and phospholipase A, activity. A membrane-bound phospholipase A, has been purified from *Escherichia coli.*

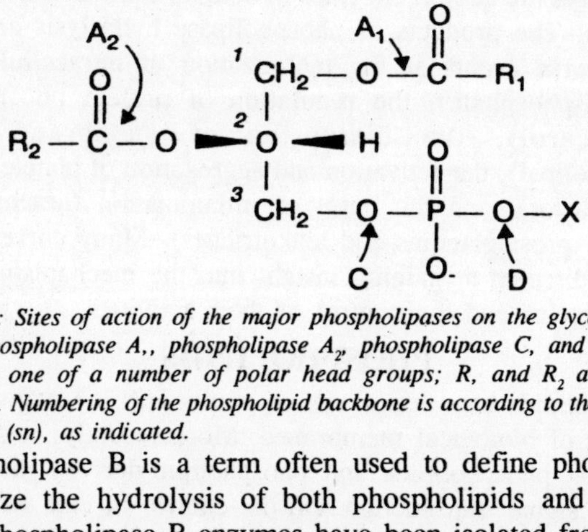

Figure 12.1 : Sites of action of the major phospholipases on the glycerol phosphate backbone: phospholipase A,, phospholipase A_2, phospholipase C, and phospholipase D. X is any one of a number of polar head groups; R, and R_2 are long chain hydrocarbons. Numbering of the phospholipid backbone is according to the stereospecific nomenclature (sn), as indicated.

Phospholipase B is a term often used to define phospholipases that catalyze the hydrolysis of both phospholipids and lysophospholipids. Phospholipase B enzymes have been isolated from bacterial and mammalian sources, but little is known about their mechanisms of action.

PHOSPHOLIPASE A_2

Phospholipases A_2 (PLA2), the most extensively studied of the phospholipases, catalyze the hydrolysis of the *sn*-2 fatty acid from phospholipids. The products of this hydrolysis are free fatty acid and lysophospholipid. The lysophospholipid product is a membrane-lytic agent and must be further catabolized to prevent injury to the cell. Lysophospholipases accomplish this feat by hydrolyzing the remaining fatty acid from lysophospholipids. Alternatively, acyltransferases can esterify another fatty acid onto the sn-2 position of the lysophospholipid to regenerate an intact phospholipid molecule. When the phospholipids contain ether linkages in the sn-1 position and are acted on by PLA_2 the lysophospholipid product is the precursor of platelet-activating factor (PAF), a cellular mediator that is implicated in many disease processes.

The fatty acid product of PLA_2 hydrolysis is also a potential cellular mediator. Arachidonic acid is a 20-carbon, unsaturated fatty acid that is primarily esterified in the *sn-2* position of mammalian phospholipids. After its release by PLA_2, free arachidonic acid can be metabolized by the cyclooxygenase or lipoxygenase enzyme system to produce prostaglandins or leukotrienes, respectively. These compounds are potent mediators of inflammation

Secretory Phospholipase A_2

Much work in the PLA_2 field has focused on secreted enzymes, since these proteins are expressed in relatively large quantities and are readily purified. Secreted forms of PLA_2 are found in snake and bee venoms and mammalian pancreatic and synovial exudates. These proteins are believed to be involved in phospholipid digestion and in the extracellular generation of inflammatory mediators.

The secretory forms of PLA_2 ($sPLA_2$) are small (-120 amino acids), water-soluble proteins that require calcium as a cofactor for catalysis. Each of these proteins is composed of a single polypeptide chain containing 10-14 cysteines, all in disulfide bonds. Hence, the $sPLA_2$s are very stable molecules. Although the $sPLA_2$s are soluble, they act in or on membranes, micelles, and other lipidwater interfaces rather than on substrates that are free in solution. Thus, the study of these enzymes or even their simple assay for diagnostic purposes requires a special understanding of the role of the interface.

The amino acid sequences of all the known $sPLA_2$s have been compared and aligned. Regions of primary structure conservation among the $sPLA_2$s include the calcium-binding loop and the histidine-aspartic acid pair at the active site. The cysteines constitute the bulk of the sequence conservation between the mammalian, reptile, and insect $sPLA_2$s. Secretory PLA_2s can be classified as group I, II, or III based on their disulfide bonding patterns, but there is also a large degree of secondary structure conservation among these proteins. As might be expected, X-ray crystallographic studies indicate that the three-dimensional structures of the secretory PLA_2s are conserved as well.

Taken together, these data indicate that the $sPLA_2$s have similar mechanisms of action. However, despite their similarities in sequence and structure, preferences for the head group type and physical form (monomer, micelle, vesicle) of the phospholipid substrate vary among the $sPLA_z$s. Many current research efforts are directed at determining the sequential and structural constraints that contribute to these differences.

Cytosolic Phospholipase A_2

Recently, several laboratories have reported the purification, characterization, and cloning of a cytosolic PLA_2 ($cPLA_2$). As predicted based on its location in the cytoplasm, this protein shows no sequence homology to the secretory PLA_zs. Cytosolic PLA_2 is larger than the $sPLA_2$s (-700 amino acids), has no disulfide linkages,

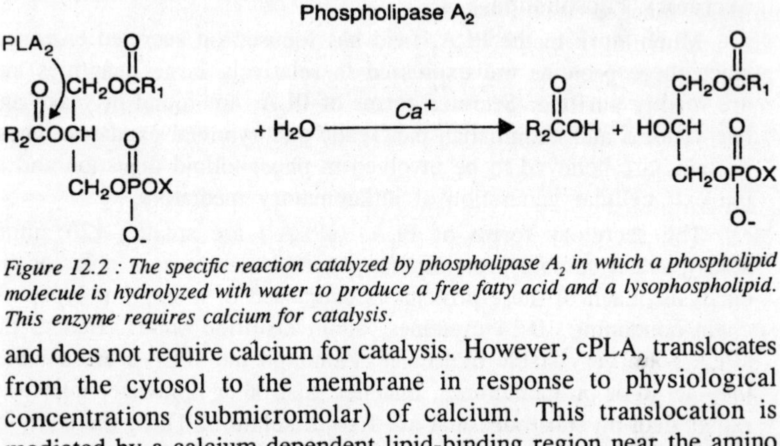

Phospholipase A_2

$$\underset{\text{PLA}_2}{R_2\overset{O}{\overset{\|}{C}}OCH(CH_2O\overset{O}{\overset{\|}{C}}R_1)(CH_2O\overset{O}{\overset{\|}{P}}OX\text{—}O^-)} + H_2O \xrightarrow{Ca^+} R_2\overset{O}{\overset{\|}{C}}OH + HOCH(CH_2O\overset{O}{\overset{\|}{C}}R_1)(CH_2O\overset{O}{\overset{\|}{P}}OX\text{—}O^-)$$

Figure 12.2 : The specific reaction catalyzed by phospholipase A_2 in which a phospholipid molecule is hydrolyzed with water to produce a free fatty acid and a lysophospholipid. This enzyme requires calcium for catalysis.

and does not require calcium for catalysis. However, $cPLA_2$ translocates from the cytosol to the membrane in response to physiological concentrations (submicromolar) of calcium. This translocation is mediated by a calcium-dependent lipid-binding region near the amino terminus of $cPLA_2$. This region has sequence homology to the isotypes of protein kinase C and phospholipase C that translocate in response to calcium.

There are several functional differences between the sPLAZs and $cPLA_2$. Unlike the $sPLA_2$s, which have no preference for the fatty acid esterified at the *sn-2* position of the phospholipid, $cPLA_2$ preferentially hydrolyzes phospholipids containing an *sn-2* arachidonic acid moiety. While the $sPLA_2$s only hydrolyze *sn-2* fatty acid from intact phospholipid, $cPLA_2$ has lysophospholipase activity as well.

Thus, $cPLA_2$ has an inherent mechanism for the catabolism of the membrane-active lysophospholipid products of the PLA_2 reaction.

PHOSPHOLIPASE C

Phospholipase C (PLC) catalyzes the hydrolytic removal of the head group from phospholipid to produce diacylglycerol (DAG) and a phosphomonoester such as inositol phosphate. While PLCs have little or no preference for the fatty acids esterified in the *sn*I and sn-2 positions of phospholipids, phosphatidylcholine (PC) preferring and phosphatidylinositol (PI) preferring PLC activities have been described in mammals, insects, and bacteria. In addition, some isoforms of PLC hydrolyze phosphatidylinositol glycans to convert lipid-anchored membrane proteins to their soluble forms.

Three isotypes of PI-specific PLC (beta, gamma, and delta) have been described in detail. Molecular weights and amino acid sequences vary widely among the PLC isoforms, but domains with considerable amino acid sequence homology have been identified and are postulated to be of functional significance. Among these is the carboxy-terminal, calcium-binding region that is believed to bind the calcium required for PLC catalysis. As indicated earlier, this region has homology to the calcium-dependent lipid-binding motifs of $cPLA_2$ and protein kinase C and thus controls PLC translocation as well.

PLC activation has been tied to the signal transduction cascades of many cell surface receptors. The mechanism of activation varies among the receptors and the PLC isoforms. The engagement of hormone receptors by ligand (e.g., vasopressin, thromboxin, angiotensin) results in G protein-mediated activation of PLC-(3. These G proteins are members of the G_q family. In contrast, the activities of the gamma isoforms of PLC are regulated by phosphorylation. These proteins are phosphorylated both by growth factor receptor kinases (e.g., EGF, PDGF, NGF) and by the nonreceptor kinases that are involved in the signal transduction cascades of T- and Bcell receptors. The interaction between the receptor kinases and PLC is believed to be mediated by the *src* homology (SH) domains of PLC—y, which are homologous to the regulatory regions of the *src* kinases.

The products of PLC activation are also involved in signal transduction cascades. When the phospholipid substrate is phosphatidylinositol-4,5 bisphosphate (PIP_Z), one product of PLC hydrolysis is inositol 1,4,5-tisphosphate (IP_3), a mediator of intracellular calcium release. Intracellular calcium fluxes control enzyme activities in many cell activation systems. Diacyglycerol, the other product of PLC hydrolysis, has been linked to the regulation of protein phosphorylation by protein kinase C. This phosphorylation, in turn, alters the activities of cell proteins.

PHOSPHOLIPASE D

To date, the phospholipases D (PLD) are the least characterized of the major phospholipases. PLD activities have been described in plants, in trypanosomes, and in a variety of mammalian cell types. Since both PLD and PLC attack the phospholipid head group to produce one water-soluble and one insoluble product, it has been difficult to distinguish between these activities. This problem is alleviated by a second activity that has been ascribed to PLD, the transfer of head groups between intact phospholipids and primary alcohols (transphosphatidylation, or base exchange). Thus, PLD activities can be characterized by the formation of phosphatidylethanol from phosphatidyicholine in the presence of ethanol.

Most of the observed PLD activities are specific for phosphatidylcholine. However, PLDs specific for the inositol phosphate bond of phosphatidylinositol glycans have been described in mammalian serum and trypanosomes. It is believed that PLD hydrolysis of the anchor of the variant surface glycoprotein of trypanosomes allows these organisms to routinely change their major antigenic determinants and thus to avoid host immune defenses.

13

RIBOZYME CHEMISTRY

Ribozymes are RNA molecules that can function to catalyze specific chemical reactions within cells, without the *obligatory* participation of proteins. Until recently it was believed that all *biochemical* reactions were catalyzed by protein enzymes. Since 1982, a number of RNA molecules have been found to be endowed with catalytic activity. For example, group I ribozymes take the form of introns, which can mediate their own excision from a self-splicing precursor RNA. Other ribozymes are derived from self-cleaving RNA structures, which are essential for the replication of viral RNA molecules. Like protein enzymes, ribozymes can fold into complex three-dimensional structures that provide specific binding sites for substrates as well as cofactors, such as metal ions. Ribozymes have been developed into highly specific enzymes that are capable of recognizing and *catalytically* cleaving targeted RNA molecules. A wide variety of applications for targeted ribozymes is under investigation. It is anticipated that ribozymes will become powerful laboratory tools for *investigating* gene function and may be developed into therapeutic agents useful for the treatment of a variety of viral and genetic diseases.

GENETIC INFORMATION

RNA is a unique *molecule* in the biological world because it serves two fundamentally distinct functions. It has long been known that RNA, like DNA, can serve as an informational molecule. For example, messenger RNA transcripts carry the information encoded by genes to cellular protein synthesis machinery. Many viruses, like HIV, use RNA as their genetic information.

Recently, investigators have found that some RNA molecules resemble protein enzymes in that they can catalyze specific *biochemical* reactions. The discovery that RNA can be a catalyst as well as a storehouse of genetic information has *revolutionised* our view of RNA and has led to the speculation that catalytic RNA may have played a critical role in the *prebiotic* evolution of *selfreplicating* systems.

CATALYTIC FUNCTION OF RNA

The catalytic activity of ribozymes results from the folded structure of RNA. Because RNA is a single-stranded molecule, its structure is very complex and is unlike the *monotonous* structure of the DNA double helix. Like protein enzymes, ribozymes fold into complex three-dimensional structures. Much is known about the secondary structure of ribozymes, which generally consists of a number of short *intramolecular* helices comprised of Watson-Crick G•C and A•U base pairs, separated by nonhelical segments. It is clear that the nonhelical segments of ribozymes are very important for catalytic activity; these *nonhelical nucleotides* interact with one another to form tertiary interactions. Because of the virtual absence of high resolution structural data on ribozymes, the details of ribozyme three-dimensional structure are generally unknown.

RNA differs from DNA in the presence of a 2'-hydroxyl group in each nucleotide. This functional group may be important for catalytic activity in three respects. First, it makes RNA much more chemically labile than DNA. Second, it participates in essential tertiary interactions within ribozymes that are essential for catalytic activity. Third, it may participate directly in the chemical reactions that take place at the active site of the ribozyme.

SELF-SPLICING RNA

The first catalytic RNA molecules discovered were self-splicing RNA molecules containing group I introns. In the absence of proteins, these *molecules* can undergo rapid and precise structural rearrangements to excise the intron and *covalently* join the exons. Group I introns contain characteristic internal nucleotide sequences and have very similar RNA folding patterns; these common structural features are essential for catalytic activity and for recognition of the correct reaction sites in the precursor RNA.

Splicing of precursor RNAs containing group I introns occurs by a two-step reaction pathway. In the first step, GTP attacks the 5'

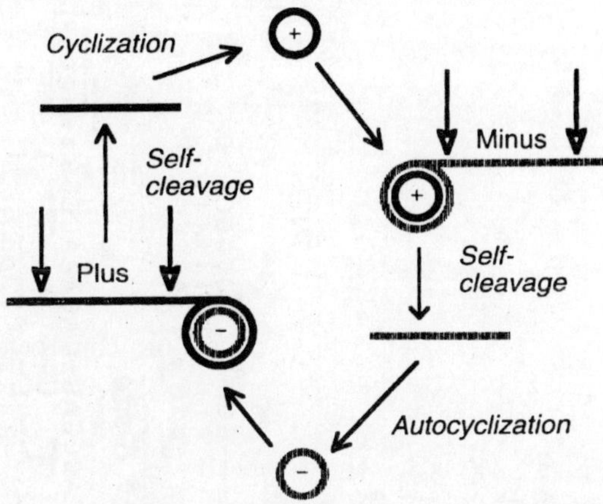

Figure 13.1 : Self-cleaving RNA in RNA replication: the replication pathway for the satellite RNA associated with tobacco ringspot virus. Rolling-circle replication gives rise to linear negative- and positive-stranded concatamers that must be cleaved and cyclized at specific sites. The self-cleaving domains in the positive and negative strands are known as the hammerhead ribozyme and the hairpin ribozyme, respectively.

splice site, becoming covalently linked to the intron and liberating the 5' exon. In the second step, the 5' exon attacks the 3' splice site, joining the exons and liberating the intron RNA. It is clear that the intron RNA itself contains the catalytic center, because it can catalyze additional reactions on internal and external RNA substrates.

A second class of self-splicing introns, the group II introns, is known. These introns share common folded structures and reaction pathways that are distinct from those of group I introns. Interestingly, group II introns appear to be more closely related to those of the nuclear pre-mRNA introns, whose splicing is catalyzed by complex nuclear particles called *spliceosomes*.

SELF-CLEAVING CATALYTIC RNA

In addition to their importance for RNA splicing, catalytic RNA molecules are essential for RNA replication in some viral systems.

For example, a circular RNA molecule associated with tobacco ringspot virus is replicated by a rolling-circle mechanism. Site-specific RNA cleavage and joining reactions are essential for RNA replication. Instead of using protein *ribonucleases* and RNA ligases to catalyze these reactions, they are carried out by the RNA *molecules* themselves. That is, during *replication* the RNA molecules contain sites at which the cleavage and joining reactions happen, as well as domains that

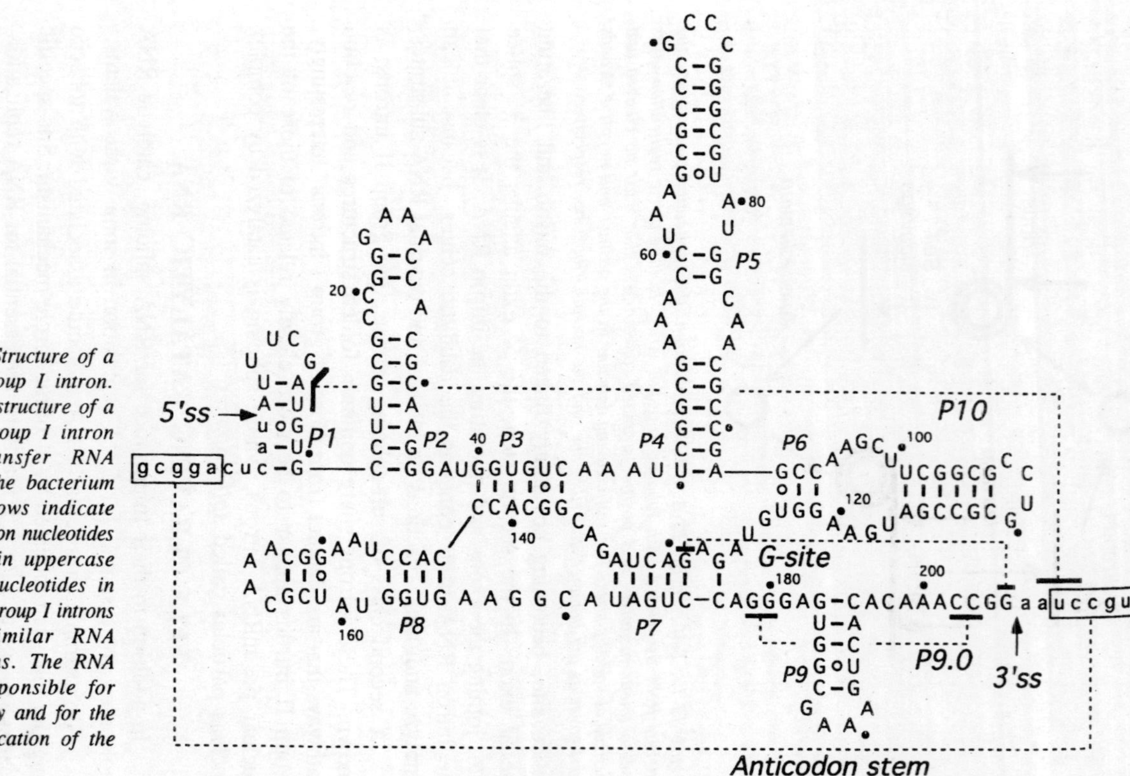

Figure 13.2 : Structure of a self-splicing group I intron. The secondary structure of a self-splicing group I intron within a transfer RNA precursor of the bacterium Azoarcus; *arrows indicate splice sites. Intron nucleotides are indicated in uppercase letters, exon nucleotides in lowercase. All group I introns have very similar RNA folding patterns. The RNA folding is responsible for catalytic activity and for the precise identification of the reaction sites.*

recognize those sites and catalyze the reactions. A similar replication pathway is followed by the RNA genome of an important human *pathogen*, hepatitis delta virus.

Deleting nonessential RNA sequences serves to isolate the catalytic domains and the sequences that are recognised and cleaved. The catalytic domains function as ribozymes to cleave the substrate in a catalytic manner; that is, each ribozyme molecule can function to cleave a large number of substrate RNA molecules. The hammerhead ribozyme is one such ribozyme derived from self-cleaving RNA molecules. The secondary structure of the ribozyme-substrate complex consists of one intramolecular duplex within the ribozyme domain

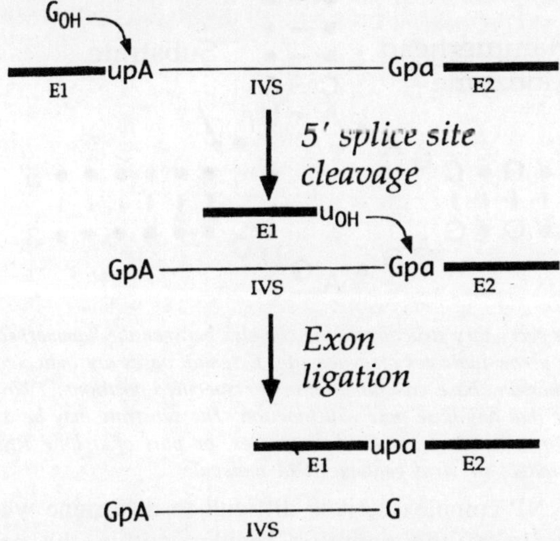

Figure 13.3 : RNA splicing pathway for group I introns. First, GTP or another guanosine attacks the 5' splice site, liberating the 5' exon and becoming covalently linked to the intron. Second, the 5' exon attacks the 3' splice site in a reaction that generates ligated exons and liberates the intron. Each of these transesterification reactions can take place in the absence of proteins.

and two intermolecular duplexes between ribozyme and substrate. Because the sequence of the intermolecular duplexes can vary in any manner that maintains base pairing, the sequence specificity of the hammerhead ribozyme can be changed by designing and synthesizing ribozymes with altered sequences in the substrate-binding arms. *Engineered ribozymes* are currently being tested for their ability to cleave targeted RNA molecules inside cells. It is hoped that ribozymes will be useful experimental tools for *investigating* the function of RNA molecules of interest. In addition, *researchers* are actively engaged in the development of therapeutic ribozymes for the treatment of serious viral and genetic diseases, for example, AIDS and cancer.

RIBONUCLEOPROTEIN ENZYME

Many cellular RNAs are bound to specific proteins, forming ribonucleoprotein (RNP) complexes. These RNPs are responsible for catalyzing a number of critical cell functions. For example, ribosomes are RNP complexes responsible for synthesizing proteins from an mRNA template. The spliceosome is the RNP complex that splices intron-containing mRNA precursors within the nucleus.

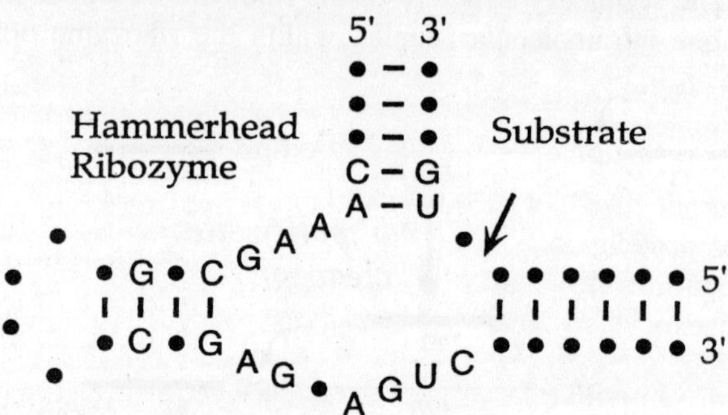

Figure 13.4 : The secondary structure of the complex between the hammerhead ribozyme and its substrate; arrow indicates cleavage site. Essential bases are indicated by letters. A dot indicates that any base can function at the specified positions. Two dots joined by a line indicate that any base pair will function. The substrate may be a short RNA oligonucleotide for laboratory biochemical studies, or part of a long RNA molecule (e.g., a cellular mRNA or viral genomic RNA molecule).

In most RNP complexes, it is difficult to determine whether the catalytic activity of the complex resides within the protein or RNAComponents. However, an accumulating body of evidence strongly suggests that RNA may be the *catalytic* component of the ribosome and the *spliceosome*.

14

Cytoplasmic Enzymes

A *cytoplasmic pool* of digestive enzyme that is the direct, or immediate, source of protein transported out of the acinar cell is a requirement of the equilibrium theory of secretion. Conversely, the cisternal packagingexocytosis theory, in all of its forms to date, excludes the existence of such a pool, namely, new protein is separated from the cytoplasm of the cell as it is being synthesized and this separation is continuously maintained by its enclosure in and transfer by means of special membrane-bound structures from the cisternal spaces of the rough-surfaced endoplasmic reticulum (RER) to the zymogen granule, and then out of the cell. Thus, the existence, or lack thereof, of a cytoplasmic pool of digestive enzyme is another clear predictive difference between the two models; one requires its presence, the other its absence. For this reason, seeking evidence for such a pool has been, and continues to be, an important focus of investigation.

Before we consider whether or not a cytoplasmic pool of digestive enzyme exists, we should ask what we mean by cytoplasm? The *cytoplasm*, as the term is commonly used today, is the liquid matrix of the cell, the solvent phase, including its contained soluble materials (dissolved solutes) and dispersed (insoluble) particles, in which intracellular structures, or organelles, such as the *nucleus* and *mitochondria*, are suspended, and which is the intracellular compartment in direct contact with the extracellular environment across the cell membrane.

This is a descriptive definition of *cytoplasm*—a description that has changed with changing times and evidence-not part of a logical system defining and subdividing the parts of a cell. For example, some insoluble material ("*dispersed particles*") is considered cytoplasmic, whereas other insoluble particles are viewed as being separate from the cytoplasm ("*intracellular structures*"). Yet both are part of a continuum of particles that is not naturally separable as a function of their particular quality. This is not to say that there are not features that distinguish one group of suspended particles from the other. Indeed, their classification is based on differences in their nature, usually differences in size, complexity, and function. However, this classification is informal and is not always consistent, and is sometimes merely arbitrary or a matter of convenience.

In general, large objects, at least in cellular terms, often, but not always, membrane-bounded, that are usually, but not necessarily, chemically heterogeneous, particularly those that are known to perform special cellular functions, such as highly organised organellar structures (for example, the *mitochondrion*), are not viewed as being part of the cytoplasm, but as being separate from it, and suspended in it. Particles that are considered part of the cytoplasm, on the other hand, tend to be smaller and less complex in terms of both structure and function; and include all soluble material, insoluble *homogeneous protein* polymers, heterogeneous aggregates of protein, and lipoprotein micelles. A useful rule of thumb in considering whether or not a particle is cytoplasmic is whether it is larger and more complex than the ribosome or smaller and less complex. That is, the ribosome forms a convenient boundary for these empirical distinctions in today's terms. Although the ribosome's chemical content and physical organisation have yet to be established fully, it is a particle of about 150A in diameter, in the main comprised of a variety of different proteins (perhaps as many as 25 or so) that appear to be designed to carry out a single type of reaction, the formation of the peptide bond.

Although one can classify intracellular elements with greater logical clarity, and this may be useful to do, the concept of cytoplasm is not part of such a classification. Rather, it is the result of a continuing experimental clarification over the years of what the cell contains. For example, before the electron microscopic era, one could successfully argue that in many cells whatever was not found in the nucleus was part of the cytoplasm. Little else was observable. Partly for this reason there are two different cytoplasms that are not necessarily congruent in practice; the anatomical cytoplasm, a geometrical space within the cell, and the biochemical cytoplasm, its molecular substance.

The anatomical cytoplasm is of course based on the appearance of the cell under the microscope. When we look at a microscopic image of a section through a cell, the cytoplasm is the material contained within the cell membrane, excluding large structures that are able to be seen at a given level of magnification. That is, the cytoplasm is the volume of the cell that is not occupied by large definable structures at a particular level of magnification and resolution. As our capability to resolve small particles has improved, the volume of the cell that is considered cytoplasmic has been reduced by their exclusion, and hence today's cytoplasm occupies a smaller percentage of the cell volume than the cytoplasm of old.

Under the *electron microscope*, the cytoplasm appears as an inhomogeneous ground substance. It excludes the interior matrices of various intracellular structures, such as the nucleus, and in practice includes all particles that are free standing, not contained within larger structures, that are as small as or smaller than the size of the ribosome, 150A diameter. Although the volume of the cell occupied by the anatomical cytoplasm may vary widely from one cell to another, we can estimate that it occupies 5080% of the cell volume in an "*average*" cell.

A full description of the biochemical cytoplasm would be a list of the nature, concentrations, physical forms, and relationships among substances contained in the anatomical cytoplasm. Ideally, the biochemical and anatomical cytoplasm should be identical in the sense that the substance of one is contained in the volume of the other; this ideal is far from being realised on the basis of current knowledge.

Our understanding of the chemical composition of the cytoplasm has in great part been derived from attempts to extract directly and collect its contents, and then to analyze them chemically. This has, for the most part, been accomplished by physically separating the cytoplasm from the rest of the cell. Usually, cells are broken open in one fashion or another (cell and tissue *homogenisation*), their contents exposed to varying degrees of *centrifugal force*, and the material that cannot be, or is not, sedimented is considered the "*cytoplasmic fraction*"; that is, the fraction of the homogenate that contains the cytoplasm.

Since this is dependent upon, among other things, the degree of applied centrifugal force, it is not surprising that as the power of our centrifuges has increased, and we have been able to sediment smaller and smaller particles, that what is included in the biochemical cytoplasm has become more and more restricted, much as with the anatomical cytoplasm. Today, materials that do not sediment readily at about 100,000 times the force of gravity are usually thought of as being cytoplasmic, although this is certainly not a firm and fast line of demarcation. Such 100,000 × g supernatants contain a large percentage of our "particle of division," the ribosome, but are most often free of substantial contamination with membrane-bound particles.

CYTOPLASMIC PROTEINS

It has been said, although not as often recently, that breaking open a cell to extract its parts, represents an act of some conceit by scientists that they can learn about the workings of an intact cell after they have destroyed its natural structure and relationships. Conceit or not, one can learn about cells in this fashion, and the effectiveness of such techniques, as well as their historical importance in the development of our knowledge of the cell, as pointed out, cannot properly be denied. From the early separation of the nucleus and with it the genetic apparatus, to the separation of the ribosome that carries out protein synthesis, this approach has been eminently rewarding. However, even if one does not think, like the vitalists of old, that breaking open the cell cannot provide any substantive information about its function, it is equally clear that studying parts of the cell isolated from their natural environment (and they can only be isolated by destroying their natural environment), must limit what we can infer about in situ events merely from knowledge of isolated cellular parts. Of course, to understand wholly cellular processes we must know the dynamical and thermodynamic relationships between parts, physical and chemical, that are extant within the whole cell.

Thus, the central problem with cell fractionation technology as we know it today, is that isolation itself limits what we can learn (at least in the absence of what one biology text calls a "*deblenderizer*"). Compounding this presently irresolvable difficulty is the fact that the most effective and common methods of cell fractionation replace the intracellular cytoplasmic environment that normally bathes the organelles with one that is very unlike it. Three major changes have commonly been made: (1) as the tissue is homogenised, the parts of the cell are diluted in a volume that is usually more than an order of magnitude greater than that within the cell; (2) the suspending medium is usually a simple, nonionic medium, unlike the cytoplasm which is chemically complex, contains relatively high concentrations of charged solutes (~150 mM), and includes a substantial concentration of protein; and (3) separation is carried out in the cold (usually 4°C), whereas cells, at least of homeotherms, are normally at 36-39°C.

There is every reason to think that these circumstances could alter, even dramatically alter, the distribution of substances between particulate and soluble phases of the homogenate, or between different particulate phases. For example, if a substance were maintained at a relatively high concentration in a particular organelle by a membrane transport process, then the dilution of the medium outside the organelle, the reduction in temperature, and altering the concentration and gradients of many other substances across the membrane of the organelle might well alter the distribution of the substance of interest. It might even lead to such substantial release from the organelle that we could mistakenly derive the impression that this substance was mainly in the cytoplasm. Conversely, in the same way, the conditions for separation might increase the access of a substance to an organelle from which it is normally excluded and we might derive the opposite erroneous impression; namely, that it is normally found in this particular structure.

To the degree that redistribution can occur, we have no ready means of determining either its occurrence, direction, or extent without an independent method of measuring the presence of the substance in different parts of the cell *in situ*. And indeed, if we had such knowledge, cell fractionation for this purpose would probably be unnecessary, since the knowledge would already be ours. Therefore, cell fractionation as a tool for determining the true intracellular distribution of substances is of limited usefulness, as *long as* the material of interest may be redistributed as a result of our experimental manipulations.

For example, one would not attempt to determine the intracellular distribution of the sodium or potassium ion by cell fractionation procedures. This would be fruitless with current technology because we would expect their distribution to be greatly altered when the cell was broken open, and in ways that we could not readily, if at all, assess. This of course does not mean that we could not study the sodium and potassium transport properties of a particular extracted subcellular structure.

However, for substances thought to be held firmly in a compartment, such as proteins within a secretion granule, held by an impermeable membrane, or contained within the membrane itself by essentially irreversible forces, separation would not alter the natural distribution of the substance as long as the organelle itself was not damaged by the procedure. Thus, if all of a particular protein were recovered in an ultraspeed supernatant, or conversely none were found there, then the conclusion that the protein is naturally contained in the cytoplasm, or restricted from it, would be drawn based on the assumption that no shift in its distribution had occurred as a result of breaking open the cell.

Most cytoplasmic and noncytoplasmic proteins have been catalogued in this fashion based on the assumption that no substantial redistribution of material occurs as a result of the separation procedure. This assumption is probably correct in many, perhaps most, cases, but is most likely incorrect in some.

In any event, it certainly is reasonable to *hypothesize,* even if we cannot rigorously demonstrate, that if all or virtually all of a material is found in the cytoplasmic fraction or is excluded from it, that this distribution reflects the state of affairs in situ. However, what are we to make of a situation in which the protein is found both in particulate and supernatant fractions of a tissue homogenate? If our assumption that no redistribution occurs is justified, then would we have to conclude that the particular protein is found naturally in both cellular compartments, and moreover, if we subdivided the particulate material and found it in a variety of subcellular sediments, then would we have to conclude additionally that it was present in all of them?

Of course this is not necessarily the case, nor is it necessarily or even usually, the conclusion that is drawn. It is commonly proposed that the material occurs naturally in one intracellular compartment, and that its presence elsewhere is an artifactual result of cell fractionation procedures. This redistribution, however, is not usually considered to be the result of a "natural" redistribution as with the sodium ion. It is usually held that such a redistribution does not occur for proteins. Rather, the presence of a protein in an "*inappropriate*" compartment is most often viewed as being the result of an "*unnatural*" redistribution; for example, due to damage to an organelle during homogenisation leading to the subsequent solubilisation of a protein component. Such a proposal of an "*unnatural*" redistribution may be bolstered by knowledge that damage to subcellular organelles may well occur during homogenisation, and in some cases clearly does.

Similarly, it may be believed that material found in a particulate fraction of a tissue homogenate is artificially derived from a natural cytoplasmic pool of this protein. In this case, it may be thought that an "*unnatural*" adsorption to the surface of the particulate material has occurred, as opposed to the protein being naturally associated with the fraction or its association being the result of a "natural" redistribution.

Nevertheless, a redistribution may either be "natural," or "*unnatural*," or even both. Nor does the occurrence of a redistribution of whatever kind necessarily mean that an organelle does not contain the material as a normal constituent as well. Of course, the point is that merely by observing a distribution of material in cell fractions, we cannot infer its causes in a rigorous fashion, that is, whether due to a "natural" redistribution (as a result of changes in gradients, temperature, and so forth), an "*unnatural*" redistribution (due to breakage or adsorption, and the like), or whether it reflects no redistribution at all. Nor, if more than one of these causes is involved, can we determine to what quantitative extent each accounts for what is observed.

These explanations of course remain hypotheses to be tested and verified in other ways. Discounting the presence of the substance in other than its primary location as being artifactual is particularly problematic. This can perhaps be appreciated by an example. Let us say that we recover most of a certain enzyme involved in oxidative metabolism in a eukaryotic cell in the particulate phase of the homogenate, and find only a minor component in the cytoplasmic or supernatant fraction. We might even know that the enzyme is concentrated in the mitochondrion, and have a good idea of how it functions chemically in concert with other enzymes in that organelle. We might also know that mitochondria were damaged to some extent during their separation from the cell. On this basis we might feel quite justified in drawing the conclusion that the small amount of this enzyme recovered in the supernatant fraction was merely an "unnatural" artifact of separation and that its sole natural home was the mitochondrion. However, such a conclusion could be incorrect. Many mitochondrial proteins are known to be synthesized on free ribosomes and released from them into the cytoplasm prior to their uptake into the mitochondrion. That is, a natural *cytoplasmic pool* of such a protein exists. Its size might be quantitatively trivial or it might be substantial, depending upon the rate of its synthesis relative to the rate of its uptake into the mitochondrion and the size of the cellular pool overall. But whatever its size, a natural pool would exist. This of course would not mean that all of this particular protein found in the high-speed supernatant would have been part of this natural intracellular pool. Some, even most, might be derived from the redistribution of the protein, either "*natural*" or "*unnatural*," from the particulate phase. Without additional information, we cannot distinguish one possibility from the others, no less quantitatively.

Thus, in considering the meaning of an observed distribution of a particular protein in fractions of cells derived from cell and tissue homogenates we need to be cognizant of the following:

1. We should not assume that an observed distribution is the natural one, the one within the cell, on the basis of the distribution alone.
2. But neither should we assume that it is not.
3. We should not assume that the presence of a protein in a particular fraction is artificial merely because this is possible.
4. We should not assume that protein is only redistributed upon homogenisation in "unnatural" ways, and that a "natural" protein redistribution cannot occur, without specific evidence for this view.
5. We should not assume that there is only a single natural location for a protein in the cell, nor that there is any particular number of natural compartments on an a priori basis.
6. We should not assume that a protein is necessarily either cytoplasmic or noncytoplasmic, because it may be both.
7. Even when the great preponderance of a protein is found in a particular fraction, we should not assume that this is the natural and sole location of the protein in the cell if we have no other evidence upon which to base our case.
8. The distribution of a molecule between fractions may change with physiological state.

SPECIFIC ACTIVITY

One way in which scientists have attempted to overcome some of the limitations of cell fractionation as a technique for determining the location of a protein in the cell has been to borrow the concept of *specific activity* from protein chemistry. Changes in the specific activity of an enzyme, that is, changes in the ratio of the activity of a particular enzyme in a sample to the total amount of protein in that sample, is used to determine the relative degree of purity of the sample and to indicate whether or not an attempt at purification has been successful.

If it has, the specific activity increases, because the quantity of other proteins contaminating the sample will have been reduced.

In applying this idea to cell fractionation, one compares the specific activity of a protein in a particular cell fraction to that found in the homogenate (that is, in the cell overall). If the particular fraction has a higher specific activity than the cell, then the inference is made that the material is concentrated in the major cell component comprising the fraction. On the other hand, a lower specific activity leads to the opposite inference, that is, the substance is a contaminant and not a natural constituent of the fraction. This approach has been useful in identifying comp[illegible]ents in which substances appear to be concentrated, particularly if the compartment occupies a small percentage of cytoplasmic volume (and therefore large differences between the specific activity of the fraction and the cell can be established).

On the other hand, this approach has its limitations. A low concentration of a substance in a compartment does not necessarily mean that it is an unnatural inclusion. It may indeed be present naturally at a low concentration. Moreover, specific activity expresses the activity of an enzyme in terms of the protein content of the fraction, and although this is a convenient measurement, it is not the proper one. If we wish to know the concentration of a substance in a particular cellular compartment, then of course we need to know the volume of that compartment. Frequently, however, the volume is unknown (or unknowable) and protein content is a convenient and useful substitute. In using it however, we make the assumption that the protein content of each and every cellular compartment is related to its volume at the same, fixed ratio, and this is undoubtedly incorrect.

INTRACELLULAR ENZYMES

Now let us look at the situation for the digestive enzymes of the pancreas. How are they distributed in the various subcellular fractions of pancreatic tissue homogenates? Is all, or almost all, of the enzyme found in the sedimentable fraction, in one vesicle or another, as the cisternal *packagingexocytosis* model proposes, or is there a "*soluble*" pool of enzyme? Using the literature as our guide, we could, on the one hand, say that almost all of the material is recovered in the sedimentable fraction. However, on the other, citing other experimental work, enzyme and animal species we could just as well say that nearly all is soluble and therefore of cytoplasmic origin. Published estimates have varied from about 85% sedimentable to about

85% in the supernatant, and most values range between these two extremes, including those in which an approximately 50-50% distribution is seen.

What sense can we possibly make of these differences? Most are not differences in the measurement of the same enzyme in the same animal species, but rather differences between different enzymes or different animals, or both. Moreover, usually the same, or similar, techniques are used to homogenize and separate the material. Therefore, the differences may not be explicable in terms of differences in methods. If we were instead considering a diverse group of intracellular enzymes, one could not conclude that all of the different proteins were either particulate or cytoplasmic, or even solely one or the other, based on such data.

However, when the secretory proteins of the pancreas have been considered, it is commonly thought that all of the various proteins should have the same in situ distribution, and therefore observed differences must reflect artifacts of redistribution and not natural variations. This view is based on the belief that the mechanism for the intracellular transport and disposition of these proteins does not distinguish between them; namely, they are handled in concert or in parallel. Thus, the same distribution should be seen for all 20 different proteins, and deviations from that distribution must perforce be artifactual, and moreover, according to the paradigm, must be artifacts of the "*unnatural*" kind, since the possibility of a "*natural*" redistribution of protein as a result of their crossing membranes is excluded a priori.

But how can we determine what is actually the real distribution and what is the *artifact*? Here the hypothesis itself often acts as the guide in a pattern of circular reasoning. The paradigm proposes that all secretory protein is excluded from the cytoplasm, and therefore all material recovered in the supernatant fraction must be artifactually derived, and studies that demonstrate in one species, or for one enzyme, or another, more material in the supernatant must be flawed, and less adequate than those that find less of a particular enzyme in this fraction. Of course, since most of these observations are derived from essentially the same cell fractionation methodology, there is no way that we can distinguish between the adequacy of one or another such study. That is, one has no clear and unbiased way of evaluating the differences and their causes. Nor can we know if they reflect an "*unnatural*" redistribution, a "*natural*" redistribution, no redistribution at all, or if material is redistributed, to what extent and as the result of what particular cause or causes. Merely on the basis of such observed subcellular distributions themselves, we have no way of determining their cause or causes, and we certainly have no right to consider arbitrarily one observation superior to another merely on the basis of our belief in a particular hypothesis.

Perhaps because of its widespread acceptance, the mere ability to imagine an artifactual explanation for an observation that runs counter to the vesicular view of secretion is often sufficient interpretive justification. That is, it is not necessary to prove that a particular artifact is responsible for what is observed. It may be sufficient merely to know that it is possible that it could be. Such an argument is bolstered, if, as is often the case, we have good reason to suppose that such an artifact occurs at least to some extent or another, although we do not know to what extent and hence do not know whether it can account for the observation. For example, one can argue that damage to enzyme-containing organelles during homogenisation or sedimentation produces an observed distribution, although one has no real knowledge that that is the case, or if it is, that it accounts for what is observed quantitatively. Or, the animal can be blamed for interspecies differences, and the conclusion drawn that one species or another is not satisfactory for studying secretion, although we have no information other than the fact that the data do not fit our hypothesis upon which to form such a judgement.

Furthermore, just as the presence of enzyme in the supernatant fraction of cells can be dismissed as being the result of an unnatural redistribution of protein from particulate fractions, one can also explain the absence of an enzyme in a particulate fraction that is derived from a cellular organelle thought to contain it, in the same way. For example, the rough endopiasmic reticulum (RER) is apparently a closed system of membranebounded spaces. It is thought that the internal or cisternal spaces of the reticulum are a major site for the storage of digestive enzyme and of secretory proteins in general. However, when one collects the microsomal fraction of homogenates, which is predominantly comprised of vesicles formed from the endoplasmic reticulum, only small amounts of digestive enzyme are found associated with it. Moreover, of the material recovered in that fraction, the great majority is either associated with the *ribosomes* or adsorbed to the surface of the membrane of the vesicle (or otherwise readily removed from it), and is apparently not within the vesicle itself. The specific activity of *digestive enzymes* in the microsomal fraction, even including ribosomal and *adsorbed enzyme,* is usually relatively low; about one-third to two-thirds of the specific activity of the whole homogenate in the rat. [N.B.: This is a low value for a large compartment (^-20% of cell volume), some 90% of which enzyme is even so not attributable to an intravesicular location.] Thus, there is very little evidence based on the fractionation of the endoplasmic reticulum that its internal spaces contain a substantial pool of digestive enzyme. Nevertheless, despite this "*negative evidence*," one can merely argue that the reason that the specific activity is low is because most of the enzyme contents of the cisternal spaces are released when the reticulum is broken open during homogenisation (that is, it is lost as a result of an "unnatural" redistribution of material). If, nonetheless, one finds little enzyme in the supernatant fraction, then we can additionally assume that the released material had been adsorbed onto the surface of particulate cell fractions (thereby also explaining the "*adsorbed*" enzyme).

In a study by Scheele et al. which attempted to decipher these relationships, a variety of complex experiments and interpretations are presented which, after all is said and done, rest on two a priori assumptions:

(1) there is no cytoplasmic pool, and

(2) there can be no natural redistribution of these proteins (namely, membranes are not permeable to them). With these assumptions in hand, they are then able to conclude from their experiments that exogenous material that associates with particulate fractions is simply due to adsorption, and that endogenous material found in the supernatant fraction is due to its "*unnatural*" release from particulate fractions.

None of this is to say that these hypotheses and interpretations are necessarily incorrect, nor that the presence of enzyme in the supernatant fraction of homogenates is proof of a cytoplasmic pool, but rather to note a pattern of reasoning in which it is thought sufficient to merely imagine an artifactual explanation for an observation that is not consistent with the paradigm in order to discount it. Of course, it is no more justified to say that the presence of digestive enzyme in the supernatant fraction is the result of an "*unnatural*" redistribution of material, than it is to say that the relative sparcity of enzyme in that fraction in another case is the result of an "*unnatural*" loss from a natural cytoplasmic pool due to adsorption to particulate fractions in the nonionic homogenisation medium. Indeed, these are not mutually exclusive occurrences. Both release and adsorption may occur, from and into a natural cytoplasmic pool, and as the result of a "natural" redistribution of material between fractions as well as an "unnatural" one. We need to be able to distinguish between these possibilities, and, moreover, in a quantitative fashion.

To do this, we must first know to what extent the contents of a natural compartment are lost, during cell fractionation, for example, from the cisternal spaces of the RER. And to know this, we must know in the first instance (in situ) what substances it contains and at what specific concentrations. Such attempts, for the cisternal spaces of the RER, have been made, but with at best unclear, if not wholly ambiguous, results thus far.

In addition to the cisternal spaces of the RER, enzyme recovered in the supernatant fraction of tissue homogenates may also be derived from the contents of *zymogen granules* broken during homogenisation. As with the endoplasmic reticulum, the question is not merely whether granule breakage occurs, but if it does, to what extent it accounts for the observed enzyme content of the supernatant fraction of the homogenate? Dandrifosse, in an interesting attempt to deal with this question, compared the number of *zymogen granules* in two pieces of tissue. In one, he measured the number of granules per unit volume in the intact cell. He homogenised the other, and then sedimented the whole homogenate and measured the number of granules per unit volume of sedimented homogenate. The number of granules per unit volume was essentially the same in both cases, and he concluded that homogenisation did not damage a substantial number of granules. Of course, this does not mean that the granules did not lose their contents as the result of a "*natural*" (permeability), as opposed to an "*unnatural*," redistribution of digestive enzyme.

The conclusion that should be drawn from my discussion to this point is not to affirm the view that cell fractionation studies are useless, but to reinforce the need to appreciate the difficulty in using this approach to determine the natural distribution of substances in the cell. Most importantly, it is hard, if not impossible, to show in this fashion that the presence of a substance in a particular fraction is wholly artifactual (or natural) unless one has prior quantitative knowledge of the natural distribution of the substance in situ in the first instance. The cell fractionation approach is often more useful when we turn the question around and ask positively and qualitatively if a substance is a natural constituent of a particular compartment to one extent or another?

PHYSIOLOGY OF DIGESTIVE ENZYMES

Our first approach, in the mid-1960s, to the question of a cytoplasmic pool using cell fractionation techniques was to look for changes in the digestive enzyme content of the supernatant fraction of tissue homogenates as a function of changing physiological state. We had observed that the gastrointestinal hormone, cholecystokinin, caused a nonparallel secretion of two enzymes; that is, it favored the secretion of one relative to the other, in particular it favored *trypsinogen* (Tg) secretion over *chymotrypsinogen* (Chtg) secretion from the isolated rabbit pancreas. At the time, we thought that this might indicate the segregation of different enzymes or mixtures of enzymes in different zymogen granules, and that we might see an inverse change in the ratio of these two enzymes in the granules remaining in the cell after stimulation. If the secretion of one enzyme was favored, then perhaps the granule pool would be preferentially depleted of this enzyme. When we performed these experiments, we found that the ratio of the two enzymes in the granule fraction had not been altered (A Tg/Chtg = 0.20 ± 0.50 SE), nor was any significant change seen in their proportions in the microsomal (RER-derived) fraction (A Tg/Chtg = 0.34 ± 0.45). The ratio was, however, changed in the postmicrosomal supernatant fraction; the fraction that contained the contents of the cytoplasm (0 Tg/Chtg = 2.35 ± 0.77). Moreover, it had changed in the predicted direction; that is, in the opposite direction to that observed in secretion, with its trypsinogen content being depleted preferentially.

Table 14.1 : Trypsinogen to Chymotrypsinogen Ratio in Subcellular Fractions of Rabbit Pancreas and in Secretion from the Rabbit Pancreas in Vitro

Source	Trypsinogen/Chymotrypsinogen
Zymogen granules	3.3 ± 0.4
Microsomes	3.6 ± 0.5
Postmicrosomal supernatant	9.2 ± 1.0
Unstimulated secretion	6.1 ± 0.4
CCK-stimulated secretion	15.8 ± 3.5

Values are means ± SE, n = 12 for cell fractions and n = 6 for secretion.

Two other aspects of the observations reinforced the view that at least a portion of the supernatant fraction's content of digestive enzyme was due to a natural cytoplasmic pool. In the first place, the ratio of trypsinogen to chymotrypsinogen in the supernatant was dissimilar from that seen in either the zymogen granule or microsomal fraction, the two major putative sources of its enzyme contents in artifactual terms . Furthermore, the enzyme ratios in both microsomal and zymogen granule fractions were similar to each other, thus excluding the possibility that the contents of the supernatant reflected a composite of loss from both of these pools. Therefore, their presence in the supernatant fraction could not have merely been due to their release from other parts of the cell as the result of an "*unnatural*" redistribution, unless there had been differential release,

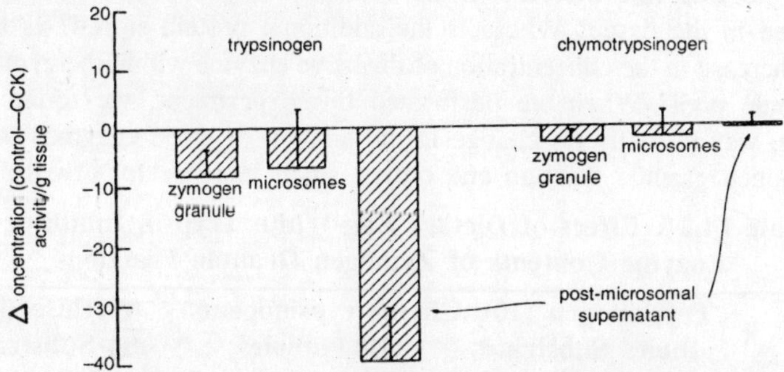

Figure 14.1 : The effect of cholecystokinin-pancreozymin (CCK) on the relative distribution of trypsinogen and chymotrypsinogen (s) in certain cell fractions from the rabbit pancreas. Values are mean ± SE, n = 8 for all groups. Animals were killed 15 min after the injection of CCK (2.0 Ivy dog units/kg body weight). Values are the differences between paired control and treated rabbits. Only the value for trypsinogen in the postmicrosomal supernatant differed from its paired control, p < 0.005.

or differential readsorption of the two proteins, or both. Only in this case could the observed proportions be explained within the context of the paradigm, that is, in the absence of a natural cytoplasmic pool. This possibility seemed unlikely because the two enzymes, *trypsinogen* and *chymotrypsinogen,* were very similar; of almost the same molecular weight, size, shape, surface charge *(isoelectric point)*, and -50% homologous in terms of amino acid composition. Therefore, extensive, *but nevertheless nonspecific,* differential release or readsorption would have had to have occurred.

In addition, the ratio of the two enzymes in secretion was quite different from that found in either the microsomal or zymogen granule cell fraction. Indeed, if anything, it more closely resembled that in the supernatant. Of course, according to the exocytosis model, the enzyme contents of secretion should closely match, indeed should be identical with, that in the zymogen granule pool.

Thus, the enzyme content of the supernatant fraction changed in a unique fashion in response to a physiological stimulus in a manner that suggested that a natural *cytoplasmic pool* of digestive enzyme was a source of the secreted material. In a second study, we tried a somewhat different approach. We produced pancreatic hypertrophy by feeding rats egg-white *trypsin* inhibitor in their food. Feeding rats, and other animals, substantial quantities of *trypsin inhibitors* produces a considerable increase in the mass of the pancreas within a few days to a week. This increase is characterised by, among other things, a two- to four-fold increase in the concentration of digestive enzymes stored in the tissue. Where is the additional protein stored? Is there an increase in the concentration of digestive enzyme within the zymogen granule pool? When we performed this experiment, we found that there was virtually no change in the amount of three enzymes in the zymogen granule fraction and only a small increase for two of the three in the microsomal fraction. Between 80 and 100% of the increase in the enzyme content of the tissue was traced to the supernatant fraction (% nonnuclear activity increase). Indeed, the specific activity of the enzymes in the supernatant was almost as high, and higher in one case, as in the zymogen granule fraction.

Table 14.2 : Effect of Dietary Egg-White Trypsin Inhibitor on Enzyme Contents of Zymogen Granule Fractions

	Trypsinogen (10) μmoles Substrate Split•min^{-1}·$gland^{-1}$	Chymotrypsinogen(s) (10), μmoles Substrate Split·min^{-1}·$gland^{-1}$	Amylase (9), mg Substrate Split· min^{-1} .$gland^{-1}$
Normal	7.6 ± 1.0	51 ± 9	758 ± 151
EWTI	7.4 ± 0.8	49 ± 6	806 ± 130

Thus, almost all of the increased concentration of digestive enzyme in the tissue was attributable to an increase in the enzyme content of the cytoplasm-containing fraction. The zymogen granules did not appear to be the source of the additional material in the supernatant, at least to any appreciable extent, because a proportional increase in its enzyme content was not observed. It was possible, however, that the cisternal spaces of the RER might have been the source of the enzyme recovered in the supernatant if virtually all of its contents had been lost during homogenisation.

To consider this possibility, we plotted the enzyme content of the supernatant versus that of the microsomal fraction from normal rats and from rats fed trypsin inhibitor. In normal rats, the two variables were correlated; that is, the more enzyme that there was in the supernatant, the more there was in the microsomes. If the endoplasmic reticulum had been the source, then we would expect the higher values seen in the supernatant fraction after trypsin inhibitor feeding to fit the same function, with data points merely being placed further away from the origin. However, this was not the case and although the relationship was maintained in that the two variables were still correlated, feeding trypsin inhibitor altered the slope of

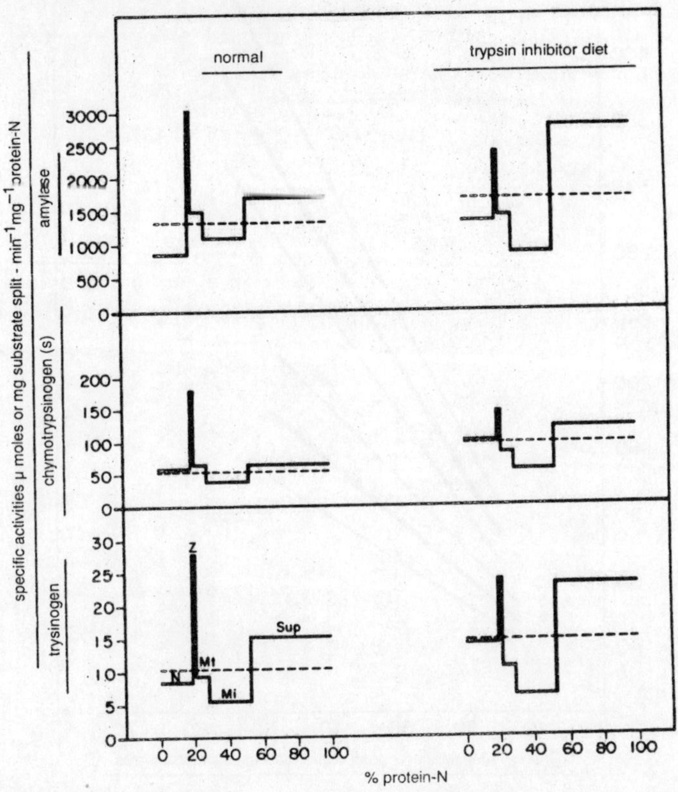

Figure 14.2 : Comparison of the specific activity of amylase, chymotrypsinogen (s), and trypsinogen in various subcellular fractions of pancreas from normal and egg white trypsin inhibitorfed rats (EWTI). (N) Nuclear fraction; (Z) zymogen granule fraction; (Mt) mitochondrial fraction; (Mi) microsomal fraction; (sup) postmicrosomal supernatant fraction. Dashed line refers to specific activity of the whole homogenate. Data are means of 10 experiments for both normal and EWTI-fed animals, except for amylase where $n = 9$.

the function so that there was considerably more enzyme in the supernatant fraction for a given amount in the microsomes. That is, feeding trypsin inhibitor led to a preferential increase (relative to zymogen granule and microsomal fractions) in the enzyme content of the supernatant fraction. Thus, the additional enzyme recovered in the supernatant appeared to be due to the cytoplasm contained therein.

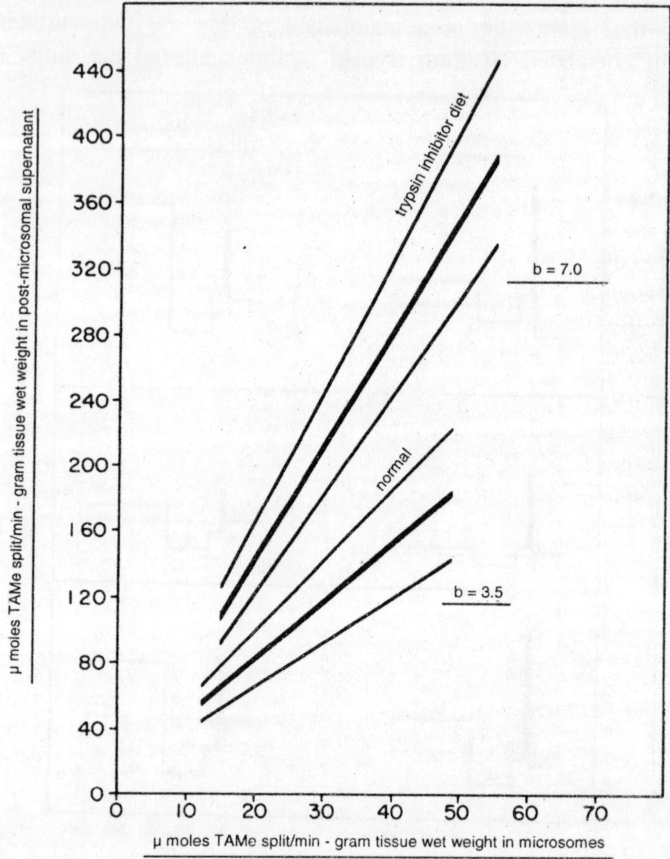

Figure 14.3 : Trypsinogen in microsomal versus postmicrosomal fractions for normal and EWTIfed rats (expressed as TAMe esterase activity per gram tissue wet weight). Goodness of fit for both lines, $p < 0.01$. Intervals around each line give 95% confidence limits. Difference between slopes, $p < 0.001$, $n = 15$ for each line.

ANATOMICAL APPROACH

Cell fractionation, the "*biochemical*" approach of determining the location of a protein in different subcellular compartments, has numerous problems. Indeed, in the final analysis, if conclusions drawn from such data are to be unambiguous, other in situ evidence is generally necessary. Investigators have been well aware of this problem, and as a result have sought other means, either independent or corroborative, of determining the location of specific proteins within the cell. A common approach has been to combine the anatomical and the biochemical by attempting to measure the chemical content of particular "*intact*" intracellular spaces in tissue sections. Two procedures have been used in pancreas. In one, the protein is labeled with an amino acid that contains the isotope tritium, and its presence over different parts of the cell determined from *radioactive emissions* emanating from the protein. This, of course, is *autoradiography*. The second approach is cytochemical and conceptually quite similar. In it a complex or precipitate is formed between an antibody to a specific protein and the protein itself (the antigen). One then determines the location of the protein by looking for the precipitate in tissue sections under the microscope. Nowadays, the antibody often is first coupled to an "*opaquing agent*" such as gold or ferritin to make the complex more easily seen in the electron microscope.

Both of these approaches have their particular strengths and weaknesses. For example, a photographic emulsion can be exposed to an isotopic emission at a substantial lateral distance from the site of its origin within tissue, and as a result the "*resolution*" of autoradiography is limited. For this reason, it has not been possible to determine whether or not a particular emitting protein is within the cisternal spaces of the endoplasmic reticulum or outside of it, or whether or not emitting proteins are contained within or outside of the small vesicles that have been proposed to account for intracellular protein transport. Although the antibody method has greater resolution, producing as it does small opaque particles at the site of precipitation, it still can present similar problems of resolution.

Furthermore, in autoradiography the period of time chosen for exposure of the emulsion to the labeled tissue, and a host of other technical details, such as the particular emulsion chosen, can have an important and often uncertain bearing on what is observed, particularly when quantitative issues are involved. Indeed, one can make grains appear or fail to appear essentially at will by these choices, especially if the emitting protein is present at a low concentration in a given pool relative to another. Moreover, the investigator, often, with perfectly good intentions, by these choices accentuates pools that contain the material at high concentrations, to make the evidence "*clearer*," by reducing the "*backround*"; namely, pools that contain the protein at a low concentration. There are similar problems with the antibody technique, which is further complicated by the fact that it can only indicate the location of the protein on the surface of a tissue section, where binding takes place, and at which site protein loss and redistribution are most likely to occur, rather than within the tissue section.

This raises an additional problem that both techniques share, that although not technically limiting, has made many studies (and certainly those that have been performed on the pancreas) uncertain and ambiguous as a means of determining the natural distribution of proteins in various parts of the cell. This is particularly so when proteins that are known to be secreted; that is, that in one way or another cross membranes are studied. The problem concerns the handling of the tissue from the time that it is exposed to the marker (antibody or isotope) to the time that it is viewed in the microscope.

To prepare labeled tissue for viewing under the microscope, the sample is often washed in some fashion or another to remove "excess" label or antibody. It is then fixed, washed again, dehydrated in organic solvents, embedded in plastic, and sectioned. To know that the material seen in the microscope bears an accurate resemblance, in terms of concentration and location, to that found normally in situ, we must know what has been lost or redistributed, and from where within the cell, during this preparation for viewing so that we can correct for it; or alternatively, we must know that nothing has been lost, or that whatever is, is either quantitatively trivial, or lost equivalently from all parts of the cell. Moreover, we cannot assume that all proteins and all tissues behave the same way in this regard, or perhaps even more significant here, that even the same protein in different compartments of the same tissue behave identically.

Perhaps when intact tissue is washed after its exposure to a marker, it is thought that no loss or redistribution of protein can occur, because such loss presupposes the existence of membrane permeability to protein. However, to assume that it cannot occur, for example, in studies on pancreatic protein secretion, is to assume that the cisternal packaging-exocytosis or vesicle model is correct while performing experiments that are presumably designed to test the validity of the hypothesis. And even this assumption of impermeability cannot be made during the preparation of thin tissue *sections* for viewing where no membrane barrier exists.

Not only does the potential for the loss of material have to be evaluated, but more difficult to deal with, loss may occur in a differential fashion from one cellular compartment to another in the same tissue. For example, in the pancreas, the digestive enzymes are found at a relatively high concentration in the zymogen granule, and moreover, in a bound state. Whereas, the same proteins in the cytoplasm may be present at a lower concentration and in soluble, not bound, form. Thus, it would not be surprising if protein was more readily lost from the cytoplasm than *zymogen granules*. This is a particularly troubling problem with the antibody technique, in which case the liability of a few molecules at the surface of the tissue section might substantially alter the apparent distribution.

Whether for the sake of editorial brevity, because the proper controls are very difficult to perform, or because they are thought unnecessary, the treatment that a tissue receives after its exposure to a marker is often not noted in detail, nor is it often reported how and to what extent these treatments affect the concentration and distribution of particular proteins within the cell. Such knowledge is, however, crucial, particularly if we are interested in quantification, and not merely in noting some compartments in which a substance appears to be found. In the absence of this evidence, we can only assume that whatever we do to the tissue subsequent to its exposure to the marker has no effect either on the concentration or distribution of the substance. And of course, such an assumption, not based on evidence, is unwarranted.

This does not mean that these techniques are useless or hopelessly inadequate otherwise. For example, it has been shown visually that radioactively labeled secretory protein is in the *zymogen granule,* and that antibodies to a variety of these proteins complex with the protein contents of the granule. The problem arises when one attempts to make quantitative estimates of the relative concentrations of substances found in various parts of the cell, particularly if one expects to be able to exclude the presence of a particular protein from a given location.

Two relatively recent studies that have used the antibody technique to evaluate the compartmental distribution of secretory proteins in the acinar cell, as examples of the problems that one faces in such studies, as well as examples of the assumptions that may underlie such attempts at compartmental analysis. In the first study, the investigators demonstrated that ferritin-labeled antibodies to several digestive enzymes bound to sections of zymogen granules. However, they reported little in the way of opaque particles over either the cytoplasm or the cisternal spaces of the endoplasmic reticulum. The presence of particles over the zymogen granules was certainly expected, but what were they to make of their absence over the cytoplasm and cisternal spaces? Their conclusions are instructive. Should the investigators conclude that digestive enzymes are not normally found in either the cytoplasm or the cisternal spaces, or should they conclude that their technique was faulty and resulted in the loss of protein from both of these compartments? They answered this question in opposite ways for the two compartments, cytoplasm and cisternal space. Since they apparently believed that digestive enzymes are normal constituents of the cisternal spaces of the RER, in accord with the paradigmatic view of things, they concluded that the absence of opaque particles over the cisternal spaces must have been due to the protein having been washed out during tissue preparation. On the other hand, for the exact same tissue, treated in exactly the same fashion, they concluded that the relative sparcity of label over the cytoplasm indicated that digestive enzymes are not normally found in this cellular compartment. Indeed, the presence of some particles over the cytoplasm, led to the conclusion that they must have been an artifactual *addition*. Of course, one cannot appropriately apply one standard of evaluation to one compartment, and another to another in the same tissue, merely because it fits one's hypothesis. In this way what begins as an attempt to test a hypothesis does nothing of the sort, and the evidence is merely used to reinforce prior notions. The plain fact is that the investigators had no way of knowing (at least as one can judge matters based on published evidence) whether the materials had been lost or not, and if so, from where and to what extent.

In the other study, gold-conjugated antibodies were used. In this case, the investigators *found* label over both the cytoplasm and the cisternal spaces. Once again the conclusions are instructive. Did they conclude that enzyme is normally found in both places, cytoplasm and cisternal spaces, disagreeing with the apparently contradictory observations of the other group? No indeed. They drew exactly the same conclusions. Label found anywhere in the basolateral region of the cell (in the cytoplasm, as well as the cisternal spaces, and in great part apparently associated either with the ribosomes or the membranes of the RER;), indicate the natural presence of digestive enzyme in the cisternae of the RER; whereas, particles over the cytoplasm are artifactual.

Thus, two studies with apparently contradictory observations led the investigators to the same conclusions; digestive enzyme is found in the cisternal spaces and excluded from the cytoplasm. Furthermore, the same observation in each study, the presence or absence of label over both cellular compartments, led to contrary conclusions depending upon which compartment was being considered. Such is the strength of paradigms.

In the latter study, the conclusion that the cytoplasm did not contain enzyme was apparently based on the fact that the concentration of particles over this space (particles per unit area) was much lower than that found over the zymogen granules, although it was only about half that over the endoplasmic reticulum-containing area of the cell. A range of concentration differences was reported for different enzymes from about a one-third cytoplasm to zymogen granule ratio to a one-sixteenth ratio. But how does this lower concentration of enzyme in the cytoplasm bear on the conclusion that there is no cytoplasmic pool? Ad lib fed rats were studied at midmorning, at which time the cell should be relatively depleted of its contents of zymogen granules (rats being nocturnal), and although the volume density of granules was not measured, from the work of others, we can estimate that it was not likely to be more than 15% of cell volume at most. The cytoplasm of the acinar cell, on the other hand, occupies about 50% of the cell's volume. Therefore, if an enzyme is three times as concentrated in the zymogen granule, distributed in a volume that is about one-third that of the cytoplasm, then approximately equal amounts of protein would be contained in both compartments. At the other extreme, a ratio of one-sixteenth, about 20% of the enzyme would be in the cytoplasm. If for the sake of estimation we assume that only small amounts of enzyme are in other particulate fractions of the cell, then the cytoplasmic pool would contain between 20-50% of total tissue enzyme according to this histochemical evidence. These numbers fit the percentage of the enzyme activity of the tissue that is recovered in the supernatant fraction of homogenates quite well. If half of the digestive enzyme associated with particulate material is not in the zymogen granule, then these values would of course be halved, or 10-25% of the total in the cytoplasmic pool; values that are also consistent with some cell fractionation data.

Thus, the observed values are not far off from what we would predict from a straightforward interpretation of cell fractionation values. We can also estimate the expected ratio in another way. From the experiments on the flow dependence of protein secretion, we can conclude, in the equilibrium model's terms, that the digestive enzyme concentration in the cytoplasm is either roughly equal to that found in secretion or is less. It does not appear to be significantly higher. Thus, the concentration of protein in secretion can be used as a maximum estimate for the concentration of digestive enzyme in a cytoplasmic pool. The concentration of protein in unstimulated secretion collected from anesthetised rats is about 10 mg/ml of fluid, or about 3-7% of the concentration of protein in the zymogen granule or equivalent to a cytoplasm to granule concentration ratio of one-tenth to one-thirtieth at most [that is, if the concentration of enzyme in the cytoplasm is the same as in secretion, and not less, and if the estimates of granule protein concentration are not too low. Thus, once again the predicted values are in the range reported with antibody cytochemistry or even lower.

There are two other points about this study. The first is that the distribution of particles raised to different enzymes in different compartments was not parallel, that is, the proportions varied from one compartment to another. Yet the evidence does not indicate the intracellular segregation of different enzymes in different vesicles, RER cisternae, and so forth. Such a measurement is not necessary, and the observation is one of state and not rate, in which case one must ask, "How did the differential distribution come about?" That is, if we take their evidence for the distribution of proteins as being quantitative, as they claim, and that no artifactual redistribution occurred, then the nonparallel transport of the various proteins must account for their observed asymmetric distribution, which in the absence of special enzymespecific or selective vesicles can only be accounted for by their independent movement by means of membrane transport.

The second point is related to the question that I raised earlier about this type of study; that is, "to what extent is protein lost during the extensive procedures required to prepare the tissue for *microscopy*?" The data that are presented indicate that about 300 particles are found over each square micron of zymogen granule area; that is, if we sum the values for the *individual enzymes*. If we estimate conservatively that this reflects the amount of digestive enzyme in a volume of the granule equal to a granule cross-section times the diameter of a single protein molecule (let us say, about 50 A), then, since the great majority of secretory proteins were estimated in this study, we can estimate as well the protein concentration of the granule based on these numbers, and knowledge of average granule volume (0.32 μm^3 in the fasted rat). When one does this, an average estimate of approximately 20,000 molecules per granule is arrived at, or about 1-2% of the chemically estimated value. We can make a similar calculation for the digestive enzyme content of the whole cell summing the particle count for all of the enzymes over all parts of the cell, and then comparing the digestive enzyme content of the cell arrived at in this fashion to that determined chemically (about 40% of the total protein content of the cell). When this is done (using a cell volume of 1000 μm^3), we find similarly that the value obtained from the particle count is less than 1% of the chemically estimated value.

We can explain this discrepancy in two ways; most of the protein is lost from the surface of the tissue during the preparation of the section, or only a small percentage of the molecules at the surface bind to the antibody. Regardless of which is the case, the measurements of particle density cannot be considered quantitative unless we know which is responsible and to what exact quantitative extent accounts for the observed discrepancies.

In the end, this anatomical approach has as many, if not more, problems than the biochemical approach of cell fractionation, particularly when our goal is to deal with quantitative issues, such as the distribution of a particular protein or group of proteins between one cell compartment and another. Perhaps these problems can be overcome, and the use of quick-freeze techniques holds some promise, but in the final analysis, we must be able to account, mole for mole, for the protein that was in the tissue initially, both in terms of loss and redistribution. In the absence of such specific knowledge, these techniques cannot be used in a truly objective fashion.

FUNCTIONAL CYTOPLASM

Thus far, we have considered anatomical and biochemical approaches to identify the digestive enzyme contents of the cytoplasm, each of which presents its own particular set of problems. The biochemical approach, cell fractionation, requires that we break open the cell, and thereby introduces the potential, unknown in quantity and character for the redistribution of protein. The anatomical approach, cytochemistry, also has the problem of redistribution, but introduces, in addition, the potential for the loss of protein during the preparation of tissue sections for viewing. There is a third approach to identify the digestive enzyme content of the cytoplasm that avoids the problems of redistribution and loss. We can call it the physiological approach, because it evaluates the question of a cytoplasmic pool of digestive enzyme in whole functioning cells and tissues, and uses physical analysis to help understand the nature of the system. We have already discussed examples of this approach in several different experimental contexts; for example, in the last chapter, I described studies in which we measured the uptake of labeled digestive enzyme into tissue in vitro, or earlier on, kinetic relationship between the disappearance of granules from the cell during augmented secretion and the output of enzyme into the duct system, as means of evaluating the existence of a cytoplasmic pool.

These were merely different experimental applications of a kinetic compartmental analysis applied to the question of a cytoplasmic pool. This physiological approach is hardly new, but it has suffered somewhat from a lack of interest recently, perhaps because of the notion that more direct approaches, such as the structural (anatomical and biochemical) ones that I have discussed, have the advantage of being more "*direct*," breaking open, as it were, the "*black box*," and are sufficient unto themselves. Of course, the physiological approach is nevertheless still necessary. Most notably, it does not require that we deal in extrapolations from the parts to the whole, since it, in the first instance, provides a description of an intact process, in much the same way that anatomy, in the first instance, provides a description of objects. As such, it catalogues the behaviour of the system, which in the end our models and hypotheses must both be derived from and consonant with. Indeed, somehow such knowledge must be ours if our models and hypotheses are to have any claim on reality whatsoever.

The cytoplasm is, in part, defined as the intracellular pool in direct contact with extracellular space across the cell membrane. It is through this intervening compartment that free-standing intracellular structures, such as the mitochondrion and nucleus, have access to the extracellular world. Thus, in this view, the cytoplasm is the middle or intervening compartment in a three-compartment system through which substances pass as they move between intracellular structures and the environment. This model has been the paradigm for the movement of molecules in and out of various intracellular spaces since the time that it became known that such spaces and structures existed and that biological membranes were semipermeable. However, as we have been discussing, significant exceptions to this three-compartment model for transport have been proposed over the years; in terms of the present discussion, for the processes of secretion (*neurotransmitters* and *proteins*) and protein transport more generally. In these cases, instead of a three-compartment system (*intracellular structure*-cytoplasm-*extracellular space*), movement is usually thought to involve only two compartments (intracellular structureextracellular space). Transport is accomplished by means of direct, although transient, connections between intracellular structures, such as secretion granules, and the extracellular space.

If a cytoplasmic pool of digestive enzyme exists, then we should be able to demonstrate that secretion occurs via a three-compartment system. Or conversely, if it does not, then a two-compartment model should suffice to describe the secretion process. A two-compartment model fails, and the evidence in hand requires the interposition of a third, presumably cytoplasmic, compartment in the secretory pathway.

Isenman and Ho, have performed another series of experiments from the descriptive perspective of compartmental analysis that bears on this question, and that came about as the result of an unexpected observation made over a decade ago concerning the kinetic character of protein secretion, and perhaps, as it turns out, of secretion processes in general. The cisternal packaging-exocytosis model, in addition to proposing that secretion results from the direct transfer of material between two compartments, is, from a compartmental perspective, a series process. That is, protein is supposed to move sequentially through a number of compartments separated by membranes; from the *ribosome* where it is made, eventually to the duct lumen. These compartments are arranged in series with each other in much the same way as resistances in an electrical series; as current flows through one to the next, so does protein move from compartment to compartment, except that in the case of proteins movement is thought to be irreversible.

If movement through a single series of compartments leads to the secretion of digestive enzyme, then if we label secreted proteins with radioactive amino acids at the site of their synthesis, we should be able subsequently to identify the labeled protein in the various putative compartments in the proposed linear sequence in a timed-ordered fashion, thereby tracing their path from ribosome to duct lumen. Evidence for such a system had been adduced, at least in part, from studies in which radioactive amino acids were applied to the tissue for a short period of time (a "*pulse*"), and the location of the label traced as a function of time in different cell fractions. If the interpretation given these "*pulse label*" studies was correct, that is, a linear sequence of membrane-bound compartments accounts for intracellular transport, then if instead of a short discontinuous "pulse" of radioactive amino acids, we added the labeled precursors to the medium bathing the gland continuously, the following should occur. In addition to the sequence of compartments being traced by the sequential appearance of labeled protein in one and then the next compartment, the approach to isotopic equilibrium (namely, reaching the maximal attainable or steadystate concentration of labeled protein in each compartment, at which time each pool would be wholly comprised of protein made subsequent to the addition of the radioactive amino acids to the medium) should also occur in the same unique sequence.

If we were to hasten the attainment of, or approach to, the steady-state or to isotopic equilibrium, by adding a stimulant of protein secretion to increase the rate of turnover of secretory protein, then this should be reflected in an increase in the rate of attainment of steady-state radioactivity in the various compartments; expressed finally in the fluid collected from the secretory duct.

However, when we performed such an experiment we found instead that there was a sharp *depression* in the specific radioactivity of protein (radioactivity per milligram of protein) in secreted fluid under these circumstances; not as we expected, a continuing, even sharper, stimulusinduced increase, as the paradigm predicted. This meant that secretion was not derived from a single precursor pool at all, but from a minimum of two intracellular pools. The stimulant had called forth secretion from an older, less labeled, pool of digestive enzyme, thereby reducing specific radioactivity, whereas in its absence, secretion occurred preferentially from a more rapidly labeled (higher specific radioactivity) intracellular pool. It was not clear from this observation alone whether both pools had direct access to the pancreatic duct, or whether one was an indirect source, as in a three-compartment model for secretion. But it did occur to us that the cytoplasm might be one of the source pools, and we felt that the zymogen granules were likely to be the other. Since the time that we made these observations it has become clear that many, if not all secretory systems display similar behaviour, being derived from at least two intracellular pools whose contents can be called forth differentially.

It was of course possible to explain this observation within paradigmatic borders, that is, in a purely vesicular context, by adding a new layer of complexity to the cisternal packaging-exocytosis concept; new processes and events could be proposed. We could say that in the unstimulated pancreas new granules are preferentially secreted as a result of their preferential movement through the cell to its apex, or preferential binding to "*exocytosis sites*" on the membrane, or both, vis a vis older granules. Moreover, since some new granules would presumably not be secreted, because old granules must come from somewhere, this preferential behaviour would either have to be lost over time, or would only apply to some newly formed granules. Beyond this, the preferential effect would be "*canceled*" by the stimulant. This of course would mean that the stimulant would have to be able, in some fashion or another, to eliminate the selective movement or binding of the new granules. However, there was no evidence of such events, although one certainly could not exclude the possibility that they might occur.

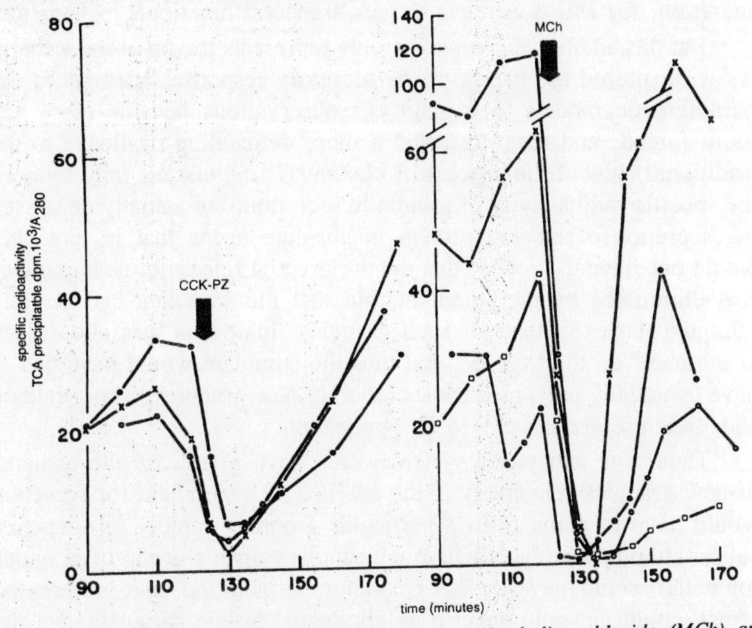

Figure 14.4 : Effect of the cholinergic agonist methacholine chloride (MCh) and cholecystokininpancreozymin (CCK-PZ) on specific radioactivity of secreted proteins. Four MCh and three CCK-PZ experiments are shown.

Alternatively, we could envision selectivity as being between cells, rather than granules, and design a parallel model of differentiated cells to account for this observation. Once again, beyond the kinetics themselves, there was no evidence that such a group of differentially responsive cells existed in the pancreas, no less that behaved in this fashion. Furthermore, in the paradigmatic view of the secretion process, the idea of a selective exocytosis of new granules seemed without clear functional meaning or purpose. Most of the material that is secreted is, even under these circumstances, derived from old granules, despite the selectivity, and very little in the way of new protein is normally secreted within an hour or so of it being manufactured either in the unstimulated or stimulated state. That is, the quantitative contribution of such a bypass would be minimal in the short run. Moreover, since this original observation, similar effects have been reported by others, not only for the pancreas, but for many other secretory cells, including neurons (for example, "*rapidly turning-over granules*" are thought by some to account for much of acetycholine release), and thus this kinetic pattern may well be a general one for secretion, for which we have to seek a general functional explanation.

But this additional complexity only represents the tip of the iceberg. As we explored the nature of the secretory response, attempts to use paradigmatic models to explain our observations became more and more forced, and thus presented a more demanding challenge to the traditional view of the process. *A cholinergic agonist* not only reduced the specific radioactivity of protein in secretion, but actually depressed the secretion of labeled protein in absolute terms that is, not only would one have to propose that the preferential release of new granules was eliminated by the stimulant, but that the stimulant had actually inhibited the exocytosis of such granules. Inhibition was about 75% at the nadir on the average, and thus the stimulant would have had to have essentially prevented the secretion of new granules, whose release had been preferred prior to its application.

Thus, our exocytosis system would have to release the contents of new granules selectively in the absence of a stimulant (or secretion would have to come from a particular group of rapidly turning-over cells), although the quantitative contribution of this group of granules (or cells) would nevertheless be minor. In addition, the preferential effect would not only have to be eliminated by the stimulant, but the exocytosis of new granules (or the secretory activity of rapidly turning-over cells) would have to have actually been inhibited.

But the complexity does not end here. An additional detail of the response to the cholinergic agonist was that the depression in the

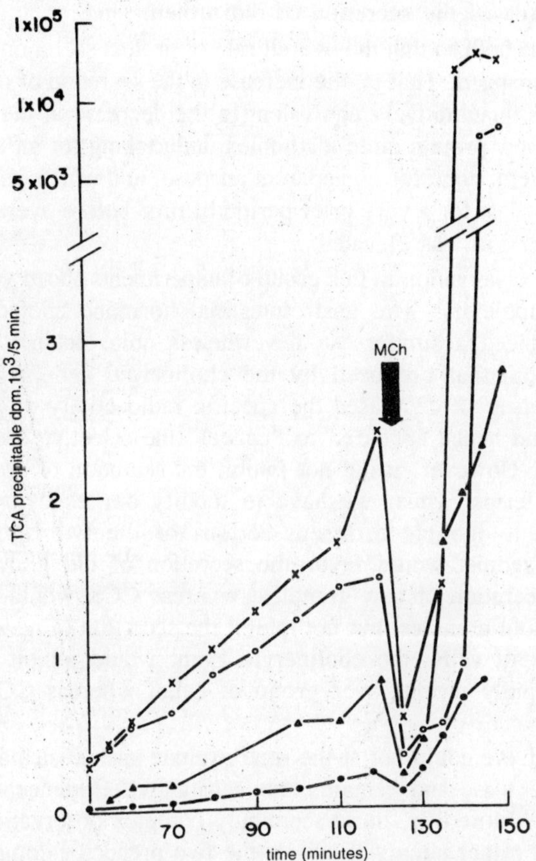

Figure 14.5 : Effect of methacholine chloride (MCh) on the amount of new (labeled) protein secreted by rabbit pancreas in short-term organ culture. Four experiments are shown.

specific radioactivity of secreted protein, as well as the absolute inhibition in the secretion of new (labeled) protein, was seen *prior* to the development of augmented secretion. That is, the stimulant produced these effects before protein secretion overall increased. How can one explain this observation in terms of the exocytosis model? At first glance this may not seem to require modification of what we have just proposed, but that is not so. Because inhibition was observed prior to the onset of augmented secretion, that is, during a period in which protein secretion remained unaltered, or at the control or unstimulated rate, the stimulant had:

(1) inhibited the secretion of new protein;

(2) increased the secretion of old protein; and

(3) done both so that the overall rate of protein secretion remained unchanged. That is, the increase in the secretion of old protein was quantitatively equivalent to the decrease in the secretion of new protein-quite a complex undertaking for an exocytosis system, done for no apparent purpose, and which effect would only last for a very brief period of time before overall protein secretion was elevated.

The last observation in this group of experiments added yet another layer of complexity. The gastrointestinal hormone cholecystokinin (CCK) produced a similar but nevertheless quite distinct effect on secretion from that produced by the cholinergic agonist. Like the cholinergic drug, it decreased the specific radioactivity of protein in secretion, and hence appeared to "cancel" the selective secretion of new protein. However, it did not inhibit the secretion of new protein in absolute terms. Thus, we have to modify our exocytosis model additionally to include different actions of the two agents. The cholinergic agonist would favor the secretion of old granules, and inhibit the secretion of new granules, whereas CCK would favor the secretion of old granules, but not inhibit the secretion of new granules (or in terms of cells, the cholinergic agent would inhibit secretion from the rapidly turning-over group of cells, whereas CCK would not).

Although we could not at the time exclude such vesicular explanations, there was, and remains, no substantive evidence that such things occur. Moreover, these seemingly complex observations could be explained rather simply if one of the two precursor compartments was the cytoplasm and if the other, in all likelihood, the zymogen granules, only had access to ductal fluid indirectly, that is, via the cytoplasm. In this case, we would merely be dealing with questions of competition between individual labeled and unlabeled protein molecules for exit from the cell; the time, nature, and extent of such competition being determined by the relative concentrations of new and older protein at the secretory site. The various effects that we observed would simply be kinetic characteristics of the secretory process, and would not have to have any particular functional meaning or purpose or require new or additional mechanisms. Competitive inhibition at the plasma membrane introduced before an overall increase in protein secretion (permeability) would, of course, inhibit the secretion of labeled protein and increase the secretion of old protein

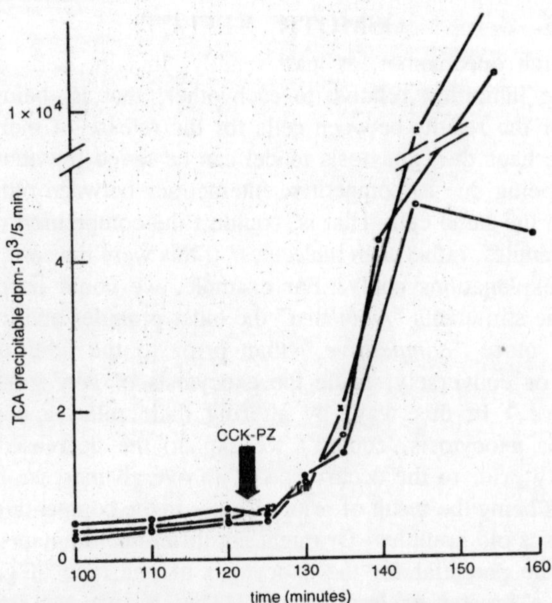

Figure 14.6 : Effect of cholecystokinin-pancreozymin (CCK-PZ) on the amount of new (labeled) protein secreted by the rabbit pancreas in short-term organ culture. Three experiments are shown.

in exact proportion, leaving the total amount secreted unaltered. This simple explanation was sufficient and no ad hoc mechanism was needed to account for this effect. The difference between the relative mix of labeled and unlabeled protein found in secretion in the unstimulated versus the stimulated state would simply be due to differences in the mix present at the apical membrane of the cell; the stimulants acting to mobilize older stored enzyme from the zymogen granules. The differences between the two hormones could simply be explained in terms of differences in the timing of the increase in secretion (*permeability*) across the apical membrane, as well as by stimulus-induced differences in the relative concentrations of enzyme from the two pools (more highly labeled and less highly labeled) at the site of secretion. Thus, although one can certainly imagine explanations for the results in terms of various complex forms of selective exocytosis, they can be explained in a simple fashion by a three-compartment model for the secretion process, in which zymogen granule contents only have access to the duct lumen by means of the cytoplasm. Furthermore, the observations in their own right were fully consistent with a three-compartment model for the secretory process, and reinforced other evidence for its existence.

OSMOTIC EFFECT

Although one cannot, at least readily, imagine cells displaying competitive inhibition relative to each other; that is, being able to account for the results between cells for the release of their enzyme content, perhaps the exocytosis model can be saved by viewing these results as being due to competitive interactions between old and new granules in the same cell. That is, couldn't the competition merely be between granules, rather than molecules? If this were the case, wouldn't the same explanations apply? For example, we could introduce the idea that the stimulants "*mobilised*" the older granules and made their exocytosis more "*competitive*," than prior to the addition of the stimulant, or conversely, made the exocytosis of new granules less "*competitive*." In this way, by altering their relative competitive position for exocytosis, couldn't we explain the decreased specific radioactivity prior to the occurrence of an overall increase in protein secretion as being the result of prior changes in the competitive position of new versus old granules? Or might not different stimulants augment or inhibit the potential for the exocytosis of one type of granule or another to a greater or lesser extent? That is, we can imagine an analogous model for competitive inhibition in which granules, rather than molecules, would be involved. And, indeed, such a possibility cannot be excluded on the basis of these observations alone, although admittedly we have to introduce the concept of specific granule mobilisation or inhibition. In any event, we attempted to distinguish between these two explanations (granule versus molecule) for the observed competitive inhibition by making use of another aspect of the secretory response to cholinergic stimulants that we had observed using the rabbit pancreas in organ culture as the experimental system.

For older protein molecules to competitively inhibit the secretion of new ones in a molecule-by-molecule fashion (to a greater extent after the addition of a stimulant, than before its addition), the concentration of old protein at the site of secretion would have to be increased, at least in relative terms, by the stimulant. Thus, the stimulant should have mobilised older protein molecules from storage pools, presumably from within zymogen granules. Since the protein inside the granule is primarily in a bound state, its release into the cytoplasm would transiently increase cytoplasmic osmolarity, at least in the apical region of the cell. Such an increase in osmolarity might in turn reduce secretory flow in one of two ways; either water would move back into the cell due to the development of an osmotic gradient, or the flux of water out in the first instance would be reduced. Although the release of protein from the zymogen granules into the cytoplasm would only produce a small increase in intracellular osmolarity, even a small change is capable of affecting secretory flow in a relatively dramatic fashion. That is, flow rate is a very sensitive osmometer, recording minor (-1%) and short-lived osmotic fluctuations within the cell.

If such an osmotic effect could be observed coincident with the competitive inhibition of labeled protein secretion, this would provide support for the view that competitive inhibition was between molecules, and that protein had been mobilised into the cytoplasm from zymogen granules; and conversely that competition did not reflect alterations in the relative rates of exocytosis of different types of *zymogen granules*.

However, although it was possible that such an osmotic effect of protein mobilisation might be seen, it was not a necessary sequelae of the release of digestive enzyme from zymogen granules into the cytoplasm. If the rate of mobilisation of protein was slow enough, then the osmotic force and the resultant changes in flow might be too small to be observed. Or if the osmoles released into the cytoplasm were accompanied by parallel increases in the release of protein into the duct lumen, thereby moving the protein and not water, then the effect also would not be seen, since no increase in osmolarity would occur. Thus, a negative observation would not be instructive.

However, we were fortunate and the cholinergic agonist that we used, at the dose that we used, produced just such a transient depression in flow from the rabbit pancreas. If one was careful in preparing bathing solutions, a reproducible, sharp, although a transient, decline in the rate of fluid flow, of about 75% at the nadir, could be produced that occurred *coincident* with the reduction in specific activity seen after the application of a cholinergic stimulant to the gland.

Thus, there was evidence that an infusion of osmoles into the cytoplasm occurred. The first question was whether or not the reduction in flow was of the order that we might expect; that is, was it quantitatively consistent with the hypothesis that it was due to the mobilisation of osmoles from the granules in the form of secretory proteins. We would expect the amount of enzyme secreted in the early phases of the response to the secretatgogue to be of roughly the same order as the number of protein osmoles liberated into the cytoplasm just prior to its secretion (as estimated from the reduction in flow). When such a calculation was made, the reduction in flow in osmotic terms and the amount of protein subsequently secreted were quantitatively quite similar.

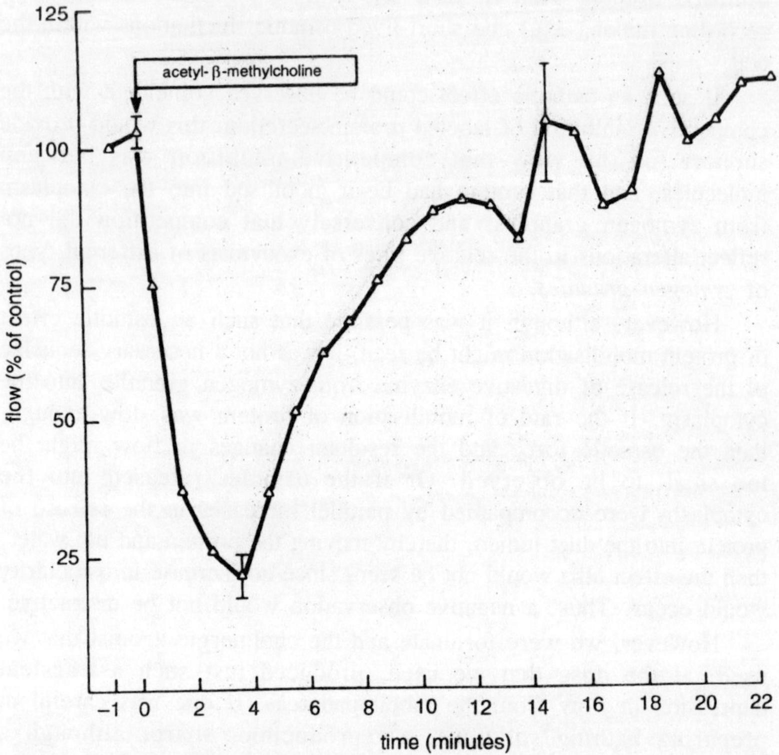

Figure 14.7 : Effect of cholinergic agonist, acetyl-p-methacholine (1.0 mg/100 ml bath volume), on flow from cannulated pancreatic duct of rabbit pancreas in short-term organ culture. Stimulus was applied at zero time (arrow). Data are expressed as percentage of control flow rate (1 min before addition of stimulant) determined from 30-sec collection periods. Points, means; error bars, SE;n=7.

Furthermore, if the reduction in flow was due to the mobilisation of protein, then we would expect that variations in the reduction in flow seen from tissue to tissue, presumably a function of the amount of protein released into the cytoplasm, would be correlated to protein secretion or output at the onset of the secretory response. That is, the higher the concentration of secretory protein in the cytoplasm, reflected in the magnitude of the flow depression, the greater the subsequent rate of protein secretion at the onset of augmented secretion (due to the elevated concentration gradient). When these two variables, the reduction in flow and the initial output of an enzyme *(amylase)*, were plotted against each other for different experiments, they were found to be highly correlated. A lesser flow reduction was accompanied by a lower amylase secretion, and a greater flow reduction indicated that amylase secretion was going to be greater. This was just as expected if the reduction in flow had been an osmotic sequelae of the mobilisation of proteins into the cytoplasm.

Further evidence for the osmotic nature of the effect was sought by setting the osmolarity of the cells slightly higher or slightly lower than normal prior to the application of the stimulant. In this case, we should be able to alter the flow depression by biasing the system in advance; an increase in osmolarity should reduce the osmotic inhibition, and an increase should accentuate it. Both effects were seen.

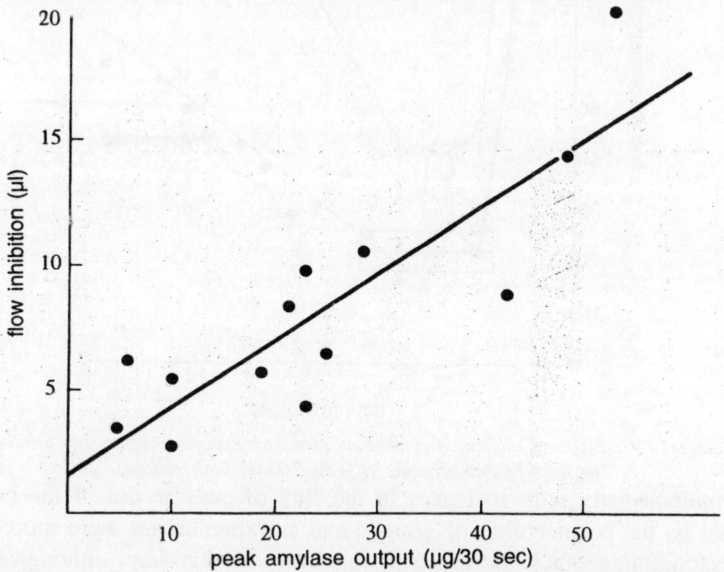

Figure 14.8 : Relationship between the magnitude of flow depression and peak amylase secretion or output produced in response to addition of acetyl-fl-methacholine (1.0 mg/ 100 ml bath volume). Total flow inhibition to nadir is plotted against peak amylase output for each individual experiment. Amylase is expressed as micrograms of crystalline a-amylase standard (730 IU/mg porcine pancreas, Worthington Chemical) per 30-sec collection period. Points were fit to the line by regression analysis ($y = 0.27x + 1.53$, $r = 0.853$, $n = 13$).

We performed one additional set of experiments considering the behaviour of the system and its consonance with our previous studies in which we measured specific radioactivity. In these earlier experiments, as I discussed above, the effects of the cholinergic agonist and the gastrointestinal hormone CCK were somewhat different in that CCK did not produce an absolute inhibition in the secretion of labeled protein, as the cholinergic agonist did. In addition, the decrease in specific radioactivity with CCK did not appear prior to the augmentation of protein secretion, but at approximately the same time. These observations suggested that CCK, unlike the cholinergic drug, increased the mobilisation of proteins from granule stores more or less

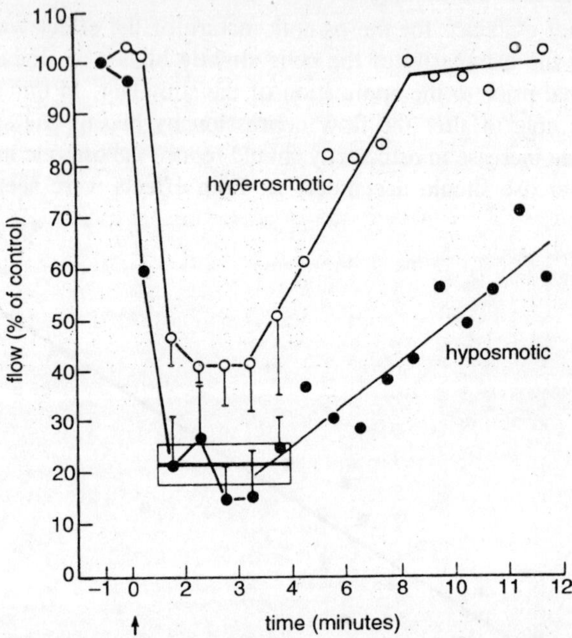

Figure 14.9 : Effect of alteration of medium osmolarity on flow depression produced by acetyl-flmethacholine (1.0 mg/100 ml bath volume).

simultaneously with increases in the flux of enzyme out of the cell; that is, the permeability of granule and cell membranes were more or less simultaneously increased in this case. In this way, although the decrease in specific radioactivity, indicating competition, would be seen, because there was also a simultaneous increase in the overall flux of protein out of the cell, there would be no absolute inhibition in labeled protein secretion. On the other hand, the cholinergic agonist would produce such an inhibition because it mobilised proteins from the granules prior to increasing protein secretion, and therefore, not only would competition occur, but there would be an absolute decrease in the amount of labeled protein secreted as well.

In light of these ideas, could we predict the osmotic behaviour that we might expect as a result of CCK stimulation of protein secretion? If it was true that both an increase in protein secretion and the mobilisation of protein from the granules occurred relatively simultaneously, then we would expect the osmotic consequence of CCK administration either to be less pronounced than seen with the cholinergic drug, or even absent if the increased concentration of granule-derived proteins was paralleled by an increase in the rate of protein secretion, so that no substantial osmotic force was generated by their release from the zymogen granule. What we found was that there was no discernible reduction in flow as a result of the addition of CCK.

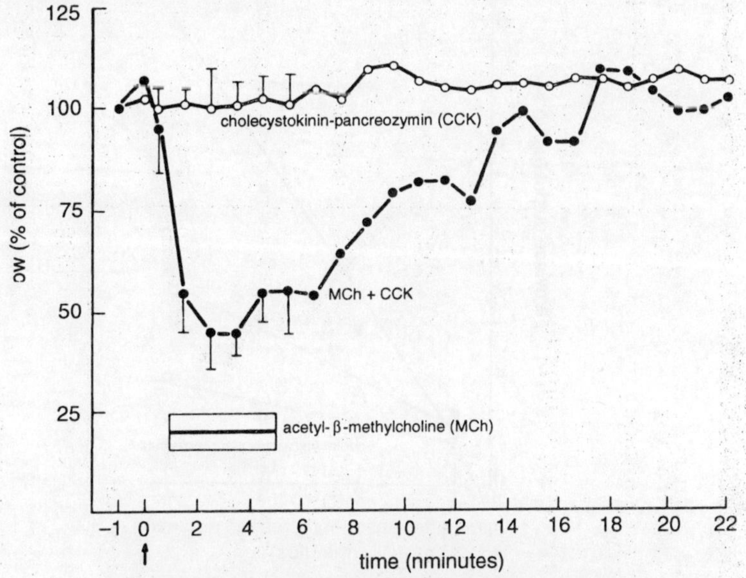

Figure 14.10 : Effect of cholecystokinin-pancreozymin (CCK, 6.25 Ivy dog units/100 ml bath volume) on flow.

If our osmotic interpretation for the absence of an effect of CCK on flow was correct, then we would expect to be able to reduce the flow depression produced by the cholinergic drug merely by adding CCK along with it. CCK would increase the release of protein across the apical membrane earlier on, and therefore would truncate the protein buildup normally produced by the cholinergic drug. When this experiment was performed, the prediction was confirmed; CCK substantially reduced the flow depression produced by the cholinergic agonist. Finally, if these interpretations were correct, then having reduced the buildup of protein in the cytoplasm in this way, the rate at which protein secretion increases as augmented secretion first begins should be less in the presence of CCK than when the cholinergic agonist is present alone. Once again the prediction was borne out by the evidence and the rate of increase of amylase secretion was greatly diminished as augmented secretion began in the presence of both CCK and the cholinergic agonist.

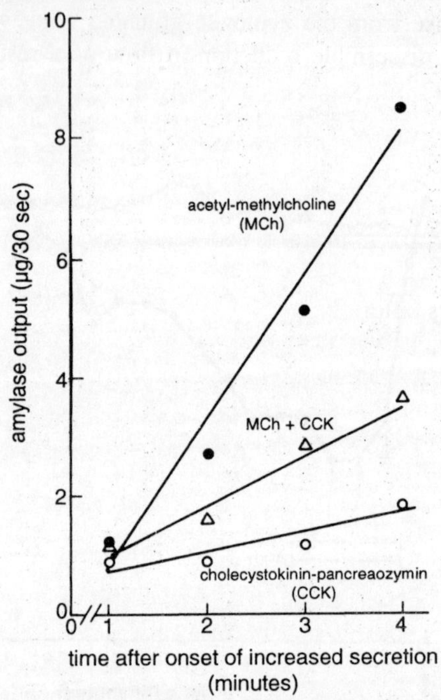

Figure 14.11 : Effect of cholecystokinin-pancreozymin (CCK, 6.25 Ivy dog units/100 ml / of bath) on rate at which amylase secretion initially increases after addition of acetyl-pmethacholine (MCh, 1.0 mg/100 ml of bath volume).

Thus, it seemed that the flow reduction that we had observed with cholinergic stimulation was indeed a sequelae of the liberation of individual protein molecules from storage pools into the cytoplasm. in conjunction with those in which specific.

15

SYNTHESIS VERSUS TRANSPORT

About 25 years after the seeming resolution of the controversy in Babkin's favor, it was demonstrated by Grossman, Greengard, and A.C. Ivy (Grossman et al., 1943) that the enzyme content of pancreatic tissue, and presumably the fluid product of the gland as well, could be altered in predictable ways by dietary means. Thus, Pavlov's view now appeared to be supported by the data; diet did affect the enzyme content of digestive fluid. However, these observations, as well as numerous substantiations in the years that followed, did not lead to the rejection of Babkin's concept of parallel secretion. Indeed, the notion was retained. The effects that were observed were attributed to changes in the relative rates of synthesis of the various enzymes by the secretory cell, rather than to their differential rates of transport out of the cell. Thus, although changes in the enzyme content of secretion did occur, in just the way that Pavlov had predicted, these changes were due to alterations in the rates of synthesis of different enzymes and not to their differential transport out of the cell from a previously formed pool in response to the contents of a particular meal.

In accordance with the parallel secretion concept, the secretion of all of the various enzymes was thought to be uniformly parallel, accurately reflecting the contents of the preexisting intracellular precursor pool of digestive enzyme. Only as the contents of that pool changed, as the result of changes in the rates of synthesis of different enzymes, would the proportions change in secreted fluid. Indeed, at all times, the enzyme composition of secretion would reflect accurately the composition of the precursor pool, and it would only vary in a dependent fashion in parallel with changes in the composition of that pool. Only when the synthesis of a new mixture of proteins had altered the composition of the precursor pool sufficiently would changes in the composition of the secreted mixture be observed.

The reasons for drawing the conclusion that synthesis was the cause of the change in the proportions of enzyme secreted was fourfold. Dietary changes led to changes in the enzyme content of the tissue. Indeed, initially all that was measured were the effects of diet on tissue enzyme content. Of course, alterations in a tissue's content of an enzyme might well reflect changes in that enzyme's rate of synthesis. Moreover, some changes in tissue content tended to follow in a sensical way the changes that had been made in the diet; for example, a high-starch diet might increase tissue amylase content. Although changes in the relative rates of transport of different enzymes might also change the composition of the cellular pool, the fact that some of the changes in tissue content were in the direction that one would expect (for a *positive feedback system*) if synthesis were responsible (in some, although not all cases, and for some, but not all, enzymes), and not in the opposite direction (which would suggest the selective transport of the enzymes out of the cell), reinforced the idea that synthesis acted to change the contents of the intracellular precursor pool. A third reason for concluding that synthesis was the cause, was that in the early studies on the subject, long periods of time (weeks to months) were used to demonstrate the effect of diet on tissue content. Such long-term effects seemed, at least on a prima facia basis, likely to reflect synthetic, rather than transport events, although the rate of turnover of digestive enzymes in the pancreas was in fact unknown, and indeed still remains to be estimated accurately in any situation, no less characterised fully as a function of state. More recent studies have made it clear that a few days generally suffice to produce substantial changes in tissue content, and such changes probably begin much sooner. Lastly, there was direct evidence that the rate of synthesis of at least some digestive enzymes could be increased by dietary means.

However, although these facts suggested the occurrence of synthetic changes, they did not demonstrate that such changes were either the primary or the sole effect of the diet, vis a vis transport, nor that changes in the contents of secretion were merely a dependent by-product of changes in synthesis. Changes in transport, particularly a comparison between the rate and extent of change in the composition of secretion, and the rate and extent of change in the proportions of different enzymes within the tissue, were not examined carefully if at all. On the basis of the available evidence, it could not be concluded that independent nonparallel transport effects did not also occur as a result of dietary manipulations. Indeed, it could not even be excluded that a transport effect was the primary, rather than a secondary, event.

This raises an important point about these studies and their interpretation. If it were true that secretion was uniformly parallel in the short run, reflecting the composition of the intracellular precursor pool, then the regulation of the digestion of single meals at the level of the digestive reactions themselves probably did not occur, because from what we know of the amount of enzyme stored within the tissue, it is previously stored material that is in great measure responsible for what is secreted subsequent to the ingestion of a particular meal, at least in many vertebrate species. Thus, these regulatory *(synthetic)* events would not in the main be for the meal undergoing digestion, but for subsequent meals, contrary to Pavlov's hypothesis. That is, in this view regulation would be predictive. If an animal is eating a high-starch meal, then the body draws the conclusion that the next meal will also be high in starch, and that it had better begin preparing. If several meals in a row were high in starch, then the effect would be reinforced, and over time the animal would adapt to a high-starch intake. This concept makes sense for grazing animals for example, who move into a field with a particular type of grass, stay there for some time, and only after having exhausted that particular food supply, move on to fresh grazing pastures, which may provide food with a different mixture of nutrients. Although such a system may certainly exist, and provide a selective advantage to the animal, for many animals and circumstances it not only would not seem to have such an advantage, but might even be disadvantageous, decreasing the efficiency of the digestion of the next meal if the prior meal was not a good predictor of the one that followed. And of course, it is even possible, in contrary creatures such as the human being, that what we eat in one meal, may predict that what we eat in the next will be different.

TRYPSIN INHIBITOR

More than fifteen years ago Wells and I performed experiments in which we considered the effects of diet on protein transport. We

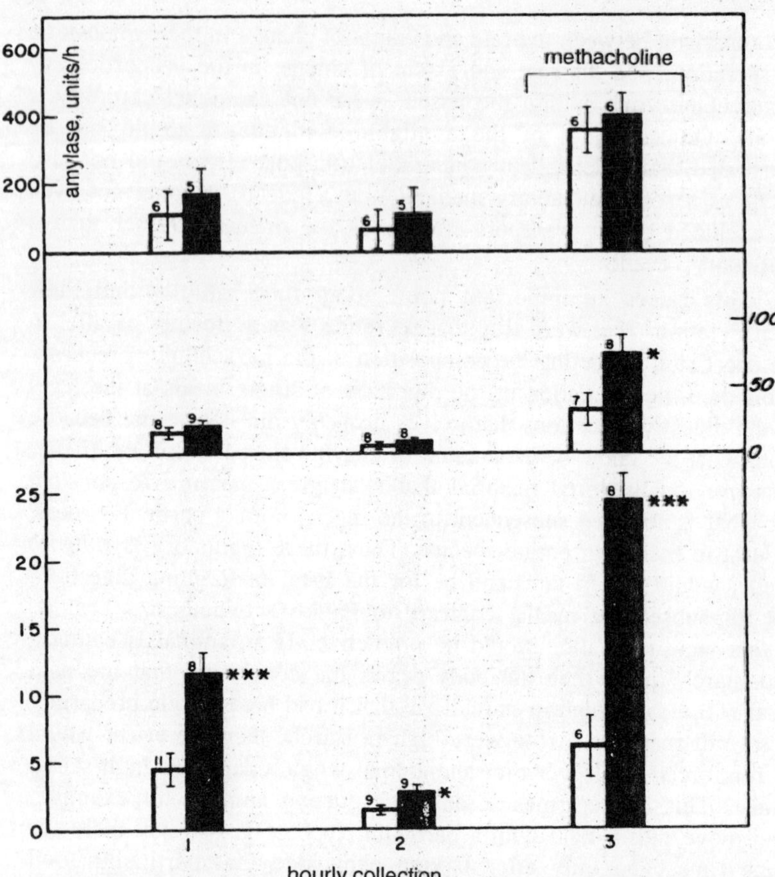

Figure 15.1 : Effect of dietary egg white trypsin inhibitor (EWTI) on pancreatic enzyme output in anesthetized rats. Enzyme outputs are expressed as units per hour.

chose a somewhat different approach from the one in which changes are made in the bulk composition of the diet, as for example, from a diet high in protein to one high in starch. We used identical, mixed diets, but added purified trypsin inhibitor (from the egg white) to the food fed one group of rats. It had been known for many years that feeding animals trypsin inhibitors produced pancreatic growth and an increase in the digestive enzyme content of the tissue. We thought that since trypsin inhibitors prevent tryptic hydrolysis, and sometimes chymotryptic proteolysis, that we might be able to see specific effects analogous to those seen when the gross composition of the diet was altered. Much to our surprise, we found that of the three enzymes

Table 15.1 : Effect of Dietary Egg-White Trypsin Inhibitor on Pancreas Weight and Enzyme Content

Diet	Body Wt (g)		Pancreas	Enzyme Content, moles or mg Substrate Split X min^{-1} $gland^{-1}$		
	Initial	Final	wt (mg)	Trypsinogen	Chymotrypsinogen (s)	Amylase
Normal (6)	215	208	832	421	3,597	6,149
4% EWTI (6)	217	209	1,104^{6}	570c	6,525d	12,976b
SEM			45	48	391	1,293
Δ%			+33	+35	+81	+111

whose presence in tissue we measured, *trypsinogen*, *chymotrypsinogen*, and amylase, the one that increased least was trypsinogen. This either meant that the effect of trypsin inhibitor was not selective, increasing the synthesis of all of the enzymes without regard to kind, or that tissue content alone was not an adequate measure of what had occurred.

We thought that the pattern of change that was observed might reflect the selective transport of enzymes out of the cell, as well as changes in protein synthesis, and attempted to evaluate this hypothesis by measuring the secretion of the same three enzymes in fluid collected from the pancreatic duct of rats fed either a normal diet or one containing the trypsin inhibitor. The results were striking. The trypsin inhibitor-containing diet did not alter the rate of secretion of amylase or chymotrypsinogen at all, but had more than doubled basal (unstimulated) trypsinogen secretion. The effect was even more pronounced when we gave a cholinergic agonist to increase the overall rate of protein secretion. In this case, trypsinogen secretion in response to the cholinergic agent was some five times greater in rats fed the trypsin inhibitor, than in controls. A lesser increase in the chymotrypsinogen response was also seen, about a doubling of the normal response to a set dose of a cholinergic agonist. However, trypsin inhibitor feeding had no effect on the extent of amylase secretion in the presence of the cholinergic agent.

These experiments were among the earliest examples of nonparallel transport that we came across. They not only demonstrated that nonparallel transport occurred, but that dietary means could be used to alter selectively the transport of different enzymes.

SYNTHETIC AND TRANSPORT EFFECTS

One can separate transport from synthetic effects by comparing changes in tissue content to changes in the secreted mixture of enzyme, or by comparing changes in the relative rates of protein synthesis to changes in the enzyme content of secretory fluid. However, the simplest and most effective method, and the one that has been most extensively used, is to look for changes in the relative rates of secretion of different enzymes over short periods of time during which little, if any, change in the enzyme composition of the intracellular precursor pool could occur. In this case, changes in the composition of the secreted mixture would reflect changes in the relative rates of release or transport of the different enzymes from a precursor pool of essentially constant composition, ergo nonparallel transport.

Of course, the greater the store of secretory product within the tissue at the start of such an experiment, the more adequately one can make this distinction, and toward this end it is useful to use brief periods of fasting (usually 12-24 hr) or other means to maximize the amount of secretory material contained within the cell at the outset of study. In addition, it is of course necessary to sample secretion over relatively short intervals of time in comparison to the time required for the contents of the storage pool to turn over. As I noted, the turnover time for the digestive enzymes in the pancreas is not accurately established, and as a result it is generally unclear at what exact rate the contents of the presecretory pool change in either stimulated or unstimulated states. Nevertheless, it is clear that in general turnover is a matter of hours, not minutes, even in the most actively secreting tissue. A fair estimate is, minimally between 5 and 20 hr, depending upon the nature of the variables. If this is so, then over a 30-min collection period, between 90 and 97.5% of what is secreted will have been manufactured prior to an imposed change of state, and changes in composition seen over this period of time cannot be attributed to new synthesis.

One can consider the measurement of turnover time from many points of view, for example, how much protein is secreted over such and such a period of time relative to the amount present in the tissue at the outset, and so forth. However, for the purpose of evaluating the cause of nonparallel secretion, the most useful approach is to determine how long it takes for new protein to find its way into secretion. If nonparallel secretion is observed before any protein manufactured subsequent to the application of the stimulant is secreted, then the effect must be due to the differential transport of previously manufactured protein. In the absence of heroic attempts to deplete storage pools, in virtually every report, in many species and under a variety of experimental and physiological circumstances, the appearance of substantial amounts of new protein in secretion is minimally about 20 min, and varies from about 20 to 75 min in most cases. In the rat, even after 3 or 4 hr of maximal secretion driven by a cholinergic agonist, and in the continuous presence of the stimulant, new protein is not found in substantial quantities in secreted fluid within an hour after the addition of a labeled amino acid to the bloodstream.

Thus, the simplest and usually the clearest method at our disposal for distinguishing synthetic from transport causes of nonparallel secretion is merely to sample secretion over relatively short intervals of time, during which we would not expect new protein to be found in secretion in substantial quantities, and from which fact we can infer that the transport of previously synthesized protein is responsible for what is observed. Most observations of nonparallel transport that have been reported involve periods of sampling, 30-60 min or less, that are generally too short to be accounted for by changes in the overall composition of the precursor pool, and occur before there is substantial secretion of new protein. Thus, the possibility that such observations of nonparallel transport can be attributed to changes in the rates of synthesis of different enzymes can reasonably be put aside. This of course, does not mean that particular treatments do not lead to the preferential synthesis of a given protein or even to the preferential release of new protein. It is just that in such circumstances the observed nonparallel transport reflects in great part (that is, quantitatively) the selective transport of previously synthesized protein, and occurs in the absence of turnover of the whole precursor pool. Of course it should be borne in mind that even if selective release of newly synthesized protein occurs, this is still nonparallel or selective transport. In any event, most if not all rapidly produced changes in the proportions of enzyme in secreted fluid appear to be primarily the result of their differential transport, and not alterations in their rates of synthesis.

PARALLEL SECRETION

Parallel transport, in which the relative proportions of different enzymes in secreted fluid reflect their proportions in the intracellular precursor pool, is a prediction of the *exocytosis model* for secretion. To the degree that the notion of parallel secretion is correct, it lends support to the idea of secretion by the exocytosis of zymogen granules. If exocytosis accounts for secretion, then we would expect the proportions of enzyme in the granule pool and in secretion to be the same or to parallel each other, the latter being directly derived from the former as a result of the bulk transfer of mass.

This is the original notion of exocytosis that includes the following subsidiary hypotheses:

(1) the enzyme contents of each granule are fixed once formed;

(2) in a given state the contents of all granules are either exactly the same or vary randomly from granule to granule to form a normal distribution;

(3) exocytosis is a probabilistic business in which a chance

association of a secretion granule and a membrane site leads to exocytosis;

(4) once exocytosis has occurred the contents of the granule are fully expelled into the extracellular environment; and

(5) neither ordered differences in the enzyme content of different granules or cells, nor the means to express any differences that might occur by chance, exist.

These nations formed the original concept of exocytosis as a completely random, mass transport process in which the proportions of different enzymes in secretion would always reflect the contents of the precursor zymogen granule pool, and stimulants or inhibitors of secretion would act by either increasing or decreasing the frequency or rate of fusion of granule to cell membrane. In each and every circumstance the proportions of enzymes in secreted fluid would remain constant or unchanged over the short run, and would accurately reflect the composition of enzyme contained in the precursor granule pool.

Thus, *parallel secretion* and *exocytosis* are conceptually congruent; evidence that transport is parallel in all circumstances is support for secretion by exocytosis and evidence for exocytosis predicts parallel transport. Therefore, testing for the parallel transport of digestive enzyme also tests the exocytosis model of secretion, at least as the idea was originally conceived. Similarly, the existence of nonparallel transport argues against the exocytosis hypothesis. In this case, we must either propose a wholly different model for secretion, modify the view of exocytosis outlined above, or consider a combination of processes. In our work, my colleagues and I have proposed, and provided evidence for, a different construct, the equilibrium theory of secretion, whereas supporters of the exocytosis model have had to modify their hypothesis to attempt to account for nonparallel transport.

PARALLEL TRANSPORT

Before we consider the data, what, in more specific terms, do we mean by parallel transport. Babkin and his colleagues appear to have had a relatively inexact definition by which they meant that there was merely a general coincidence in *responsiveness* in the secretion of different enzymes to all secretagogues. That is, all stimulants increase the secretion of all enzymes, although not necessarily in exact proportionate correspondence. When the idea of parallel transport is placed within the context of a mechanism for secretion, such as exocytosis, we can be more exact about what we mean and about what would be required. Thus, for the view of exocytosis that I outlined above, the secretion of a pair of enzymes should always covary, and increases in the secretion of one should be accompanied by coincident proportionate increases in the other under all circumstances of time, state, and rate. If we compare the rate of secretion of two enzymes, then we would expect, no matter what the circumstances or how the various samples were obtained, that a plot of the values for a variety of individual samples would form a straight line intersecting the axes at the origin. Thus, whenever the overall rate of secretion is increased or decreased, so would the rate of secretion of the two enzymes and in direct proportion to each other. As the overall rate of secretion approaches zero, so equally should the secretion of the two test molecules, and hence the line should regress through the origin; when one is not secreted, neither is the other. Furthermore, if we account for all measurement error and animal-to-animal variability in the proportions of the enzymes, then all of the points should fit the calculated regression line perfectly, that is, with a correlation coefficient of 1.0. To the extent that measurement error and animal-to-animal variability exist, the value should be less than 1.0, but whatever the correlation coefficient may be, it should be an invariant characteristic for the pair of enzymes that reflects the constancy of the secretory process and the sources of variation. That is to say, whatever the goodness of fit of the points to the line, it should be constant from circumstance to circumstance as long as the assay and the prior state of the tissue is not changed. Thus, transport is parallel whenever the relationship between the secretion of two or more proteins is constant despite changes in state or circumstance otherwise, and regardless of the overall rate of protein secretion. As a result, when the amount of one enzyme in a sample of secreted fluid is plotted against that of another, for the set of any and all such samples of secretion, a linear function should be formed that would both regress through the origin and have a constant variance from situation to situation that would be a particular and fixed characteristic of the specific enzyme pair.

LABORATRY OPINION

My first attempt in the mid 1960s to evaluate the question of parallel secretion, produced an example of nonparallel transport. It was measured in pure pancreatic fluid collected directly from the pancreatic duct in a defined in vitro experimental setting (the organ preparation of rabbit pancreas). Such reports had in the past often been made on collected duodenal chyme, and, as a result, the question of whether or not the enzymes had actually been secreted in a nonparallel fashion, or whether their activity had merely been differentially altered by other substances in duodenal fluid after their secretion, had often been a point of uncertainty. Moreover, the measurements had always been made in situ, usually in awake subjects; thus, there was a great deal of variability from trial to trial and experiment to experiment. Our report was notable because trialto-trial variability had been greatly reduced, although not eliminated, by the in vitro system; and of course, it was clear that whatever changes were observed were directly attributable to the fluid secreted by the cells of the gland.

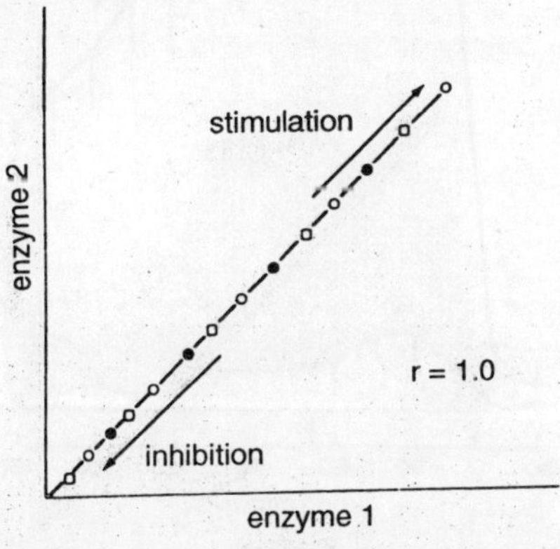

Figure 15.2 : The relationship between the secretion of two enzymes predicted by a model that proposes the obligatorily parallel or linked secretion of different secretory proteins, as in the exocytosis of zymogen granules of mixed enzyme content.

However, the particular effect itself was somewhat surprising. In the first instance, nonparallel transport occurred as part of the response to the administration of a secretagogue, the gastrointestinal hormone cholecystokinin (CCK), that produced a general augmentation of protein secretion by the gland, and not in response to the presence of a particular foodstuff or product of digestion. It was not clear why this hormone should favor the secretion of one enzyme or another on regulatory grounds, unless it represented one of a series of such hormones, and there was no evidence for this at the time. Thus, it seemed that nonparallel transport did not necessarily have to be related to the regulation of digestion. It could be a consequence, even an apparently purposeless consequence, of the administration of a general stimulant of protein secretion. Perhaps, it was in some fashion, simply a by-product of the mechanism of secretion in which differences in the "*permeability*" of the cell to different molecules or differences in

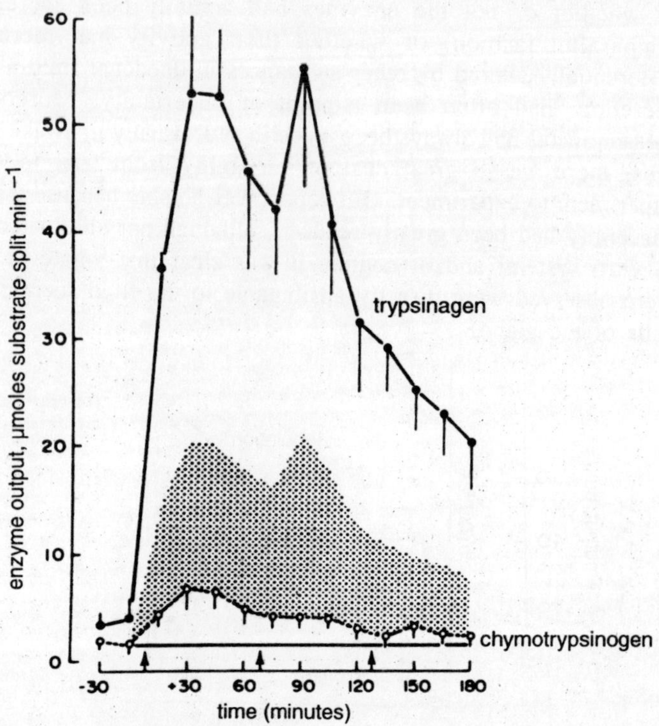

Figure 15.3 : Trypsinogen and chymotrypsinogen secretion as collected from the cannulated duct of rabbit pancreas in organ culture in response to cholecystokinin-pancreozymin (CCK PZ). Values are means ± SE for 16 preparations.

their available intracellular concentration produced by the hormone, led to their differential release.

Furthermore, the differences in the rates of secretion of different proteins were not absolute. That is, CCK did not increase the secretion of one protein while leaving that of another totally unaltered. Rather, it increased the secretion of both, but not in a proportional manner. Thus, an agent that produces nonparallel transport need not do so in an absolute fashion, augmenting the secretion of one enzyme, but not another. It might alter their rates of secretion *relative* to each other. The meaning of this, both in regard to the mechanism of secretion, and how secretion appears to be regulated, will be developed further below.

The particular observation of nonparallel transport produced by CCK in the rabbit was a disproportionate increase in the secretion of two proteolytic enzymes, trypsinogen and chymotrypsinogen, of very

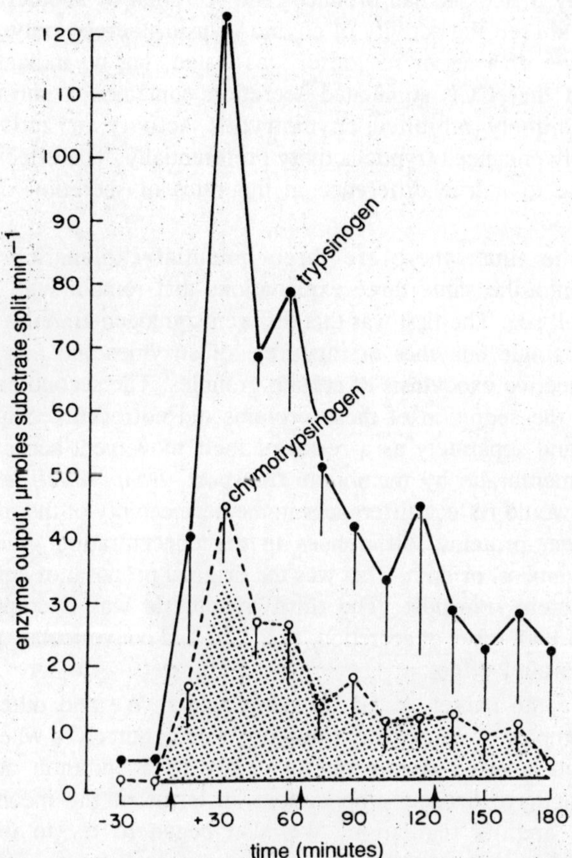

Figure 15.4 : Trypsinogen and chymotrypsinogen secretion from rabbit pancreas in organ culture in response to acetyl-fl-methylcholine. Values are means ± SE for seven preparations.

similar molecular weight, surface charge, shape, and amino acid sequence. It increased trypsinogen secretion to a proportionately greater degree, even though the concentrations of both enzymes in secreted fluid were elevated. (I have already touched on this particular observation in reference to the cytoplasmic pool of secretory protein.) On the other hand, when a cholinergic agonist, which also increased overall protein secretion, was tested instead of CCK, the increase in the secretion of both proteins was more or less parallel. Thus, using the same biological system, experimental protocol and procedures, measurement techniques, and assays, one stimulant produced the secretion of a quantitatively different mixture of proteins than the other. The possibility that CCK had produced the secretion of another substance that had altered the activity of chymo trypsin, thereby only giving the appearance of a lesser response, was tested, but no indication could be found that CCK-stimulated secretion contained a substance that either uniquely inhibited chymotryptic activity or activation, or conversely enhanced tryptic activity preferentially. The effect appeared to be due to a true difference in the rates of secretion of the two proteins.

At the time, there are three potential explanations for this phenomenon-the same three explanations that remain with us today, as we shall see. The first was that different zymogen granules contained different single enzymes or mixtures, of enzymes and that CCK led to the selective exocytosis of certain granules. The second explanation was that the secretion of these proteins did not occur en masse, but directly and separately as a result of their movement across cell and granule membrane by membrane transport. *Nonparallel secretion* in this case would reflect differences in the permeability of the membranes to different proteins, differences in the concentration gradients for their movement, or both. This was the original proposal of equilibrium-based protein transport. The third hypothesis was a composite-the mixing of both types of secretion, vesicular and nonvesicular, producing the nonparallel effect.

There the matter rested for many years. We and others sought other examples of nonparallel transport, asking ourselves whether what we had discovered was a rare anomaly or a common occurrence that might have broader implications in terms of the mechanism of secretion and its regulation. We also began to try to distinguish between different models of secretion, as well as to define such models and the deductive predictions that one might make about them, in a more specific fashion. This led to a variety of experimental strategies and approaches. Those who did not believe in the existence of nonparallel transport either simply ignored the observation or discounted it, assuming it to be some sort of assay artifact, although no coherent explanation of how this could have been was put forth.

Even though almost a decade later no satisfactory alternative explanation had been proposed, at that time three groups felt the need to attempt to "*reproduce*" the original observation. In one study the investigators reported that they were unable to reproduce the effect, and concluded that the original observation must have been due to general problems with the activation of chymotrypsinogen although

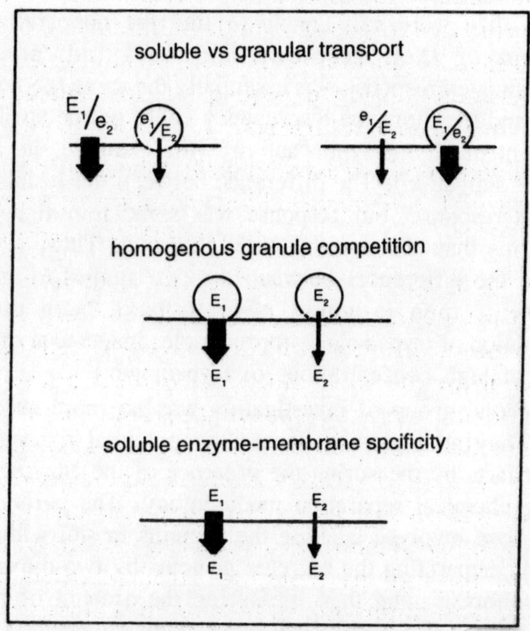

Figure 15.5 : Some transport pathways that could produce "enzyme-specific secretion" by pancreatic acinar cells.

they did not explain how such a problem could have produced what was observed (particularly, the "*better activation*" with a cholinergic agonist as the secretory stimulant). Nor, more generally, did they explain why the extent of activation or level of activity was even relevant, as long as the measurements were made in a reproducible fashion and compared with each other for different conditions by means of standard curves or functions, as was of course the case. Beyond this, these investigators had chosen a dose of CCK that was ineffective, that is, it did not significantly augment protein secretion at all. Why they thought this a more satisfactory circumstance in which to seek the nonparallel effect of CCK than an effective dose has never been made clear, and it certainly seems as if they were not looking for the response under the most propitious circumstances. In later work however, they did increase the dose of CCK, and were able to produce a modest, although by our experience, poor secretory response to the hormone. They produced an average peak increase in enzyme secretion of about threefold, and were only able to maintain this augmented secretion for about 1 hr, whereas in our hands the response lasted at least 3 hr, and the peak was severalfold greater. In any event, they were still unable to find the nonparallel effect. A comparison of their response and ours indicated that the chymotrypsinogen response was essentially the same (about a threefold increase), and therefore the discrepancy could not be attributed to the measurement of chymotryptic activity (for example, its autolysis or "*incorrect*" activation). The difference between the results was in the trypsinogen response; our response was some four times greater in relative terms than the one that they observed. Thus, if one wished to attribute the differences between the two studies to an artifact of measurement, then it would have to have been due to their underestimation of trypsinogen (for example, inadequate or incomplete activation of high concentrations of trypsinogen).

The second group of investigators was no more successful than the first. They attempted to bypass the potential for activation and activity artifacts by measuring the presence of the enzyme in secreted fluid using chemical separation methodology. The particular method that they chose involved labeling the proteins in situ with radioactive amino acids, separating the secreted proteins by two-dimensional disc gel electrophoresis, and then measuring the amount of radioactivity associated with each identified spot. Although this method sidestepped the issues associated with activity measurements, it had its own problems. Was it truly a satisfactory quantitative analysis? As a method, was it on a par, no less superior, to the very sensitive and reproducible measurement of enzyme activity? This question has yet to be satisfactorily answered, and hence it is not clear that it did not reduce the ability to make quantitative distinctions rather than enhance them.

Another problem, of which the authors were aware, was the need to label the whole storage pool prior to eliciting augmented secretion,

Table 15.2 : Secretion of Trypsinogen and Chymotrypsinogen by Rabbit Pancreas in Vitro in Response to Presence of CCK-PZ°

	Steer and Manabeb		Rothman and Wilkingc	
	1st Hr	2ndHr	1st Hr	2nd Hr
Trypsinogen	314	211	1290	1165
Chymotrypsinogen	317	183	343	246

so that the labeled protein properly traced the movement of protein overall. That is, it was necessary to be sure that tracer kinetics applied. In its absence one could only assume (incorrectly at that) that there was no difference in the movement of stored and newly manufactured protein, and this would of course in a sense be to assume the parallel transport hypothesis correct even as one was testing it. To overcome this problem, these investigators incubated the gland in vitro for many hours (about 7) prior to the application of the stimulant, although they apparently did not determine that labeling was at the steady state and that the old pool had actually turned over completely prior to the application of the stimulant. In any event, 7 hr in vitro is a long time for the biological preparation of pancreas that was used, which is generally on the edge of the useful, as seen by both its diminished secretory rate and responsiveness, by this time. Whatever problems there may have been with their approach, they were nevertheless once again unable to reproduce our result, although they did observe and report other examples of nonparallel transport.

Of course, it was possible that all of these reports were correct and that the differences between our observations and the others were real, for whatever reason, and attributable to natural causes, and not measurement problems. As the renowned physiologist Walter Cannon pointed out, it is not wise to assume that observations that you cannot reproduce are necessarily incorrect. We cannot always reproduce the exact conditions that others have applied. The expectation of these workers was no doubt that the change in ratio should be observed whatever the overall rate of secretion by the tissue might be. However, this need not be the case. Indeed, it is not the case. The function relating the secretion of *trypsinogen* to *chymotrypsinogen* in unstimulated conditions was nonlinear with a positive intercept (that is, trypsinogen secretion occurs, at least theoretically, in the absence of chymotrypsinogen *secretion)*. If, in the simplest scenario, CCK merely moved points out along this curve, not changing its parameters, then we would find that close to the yaxis intercept the ratio would be high, decreasing as the rate of secretion was elevated, and subsequently increasing again as output or secretion values approached the apparent asymptote. Thus, CCK could, by increasing the rate of secretion, reduce or leave the ratio unaltered, as well as increase it, depending upon where our particular data points fit on this function. The average value for CCK-stimulated secretion found in our original experiments is shown in Figure elsewhere in this chapter by the closed circle. It is beyond the limit of outputs observed in the unstimulated

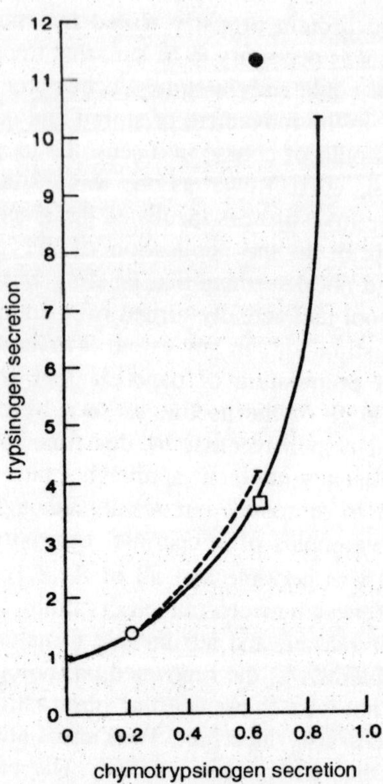

Figure 15.6 : The relationship between unstimulated trypsinogen and chymotrypsinogen secretion from the rabbit pancreas in short-term organ culture (solid line). The end of the function indicates the limit of observed values.

state (the end of the function as drawn) and displaced somewhat towards the trypsinogen axis, suggesting that the function's parameters had been altered, at least the apparent asymptote. We can look at the results of Steer and Manabe superimposed on this function. They produced roughly the same chymotrypsinogen response that we had, that is, a threefold increase over the unstimulated rate. If we assume that the control rate of secretion was approximately the same in both series of experiments (theirs and ours), then the open square estimates the enhancement that they observed, an approximately threefold increase over the control rate (open circle) if the points merely moved out the control function. In this case, the ratio would have decreased slightly in the presence of CCK from about 6.5 to 5.5. If the function was displaced somewhat toward the trypsinogen axis (as shown by the dashed line), then the ratio would have been unchanged.

Of course, this is the reason for the differences in the observations, but it does at least offer an explanation that does not fall back upon the easy "complaint" that someone else's observations are flawed or erroneous, merely because you are unable to reproduce them. Moreover and more importantly, this analysis initiates us to the fact that we cannot assume that the relationship between the rate of secretion of different enzymes is merely due to changes in the slope of linear functions (as in Figure 2) that regress through the origin and that the question of nonparallel transport has, not surprisingly, its own descriptive intricacies, which we shall consider further below.

There is another possibility to be considered in evaluating these observations in comparison with the original ones. The preparation of CCK that we used in the mid 1960s was very impure relative to material available in the mid 1970s. Our CCK preparation, the purest generally available at the time, only contained CCK as some 0.2% of its total peptides, whereas the new (1970s) material contained 20% or more as CCK-a big difference in purity! Therefore, it was possible that our effect was due to a contaminant of CCK preparations that had been purged during the intervening years. It was with this thought in mind, as well as simply the desire to see if we could repeat our own observation (breaking the scientist's adage that one *never* attempts to repeat one's own experiment), that we, the third group, undertook to reproduce our observation with the new, purer CCK preparations that had become available. We performed a substantial series of experiments, collecting secretion every 15 min for 3 hr in the presence of the hormone, thereby providing us with a fair sampling of the situation.

We found, much to our delight, that even after 10 years the effect was still there, at least in our hands, and indeed was seen with even greater clarity than before. Trypsinogen secretion was augmented to a much greater extent than chymotrypsinogen secretion and the effect lasted for essentially the whole 3-hr sampling period. Indeed, augmented trypsinogen secretion continued over a 3-hr period, even as chymotrypsinogen secretion fell and became indistinguishable from control rates. We compared our activation and assay procedures for chymotrypsinogen to those used by others and found that the differences had no bearing on the results. In addition, we obtained an even purer CCK from Professor Mutt at Karolinska Institute and found that the purest available material (called "pure") also produced the nonparallel effect. *Cholinergic* stimulation was also tested again and found again not to produce nonparallel transport from this organ preparation for

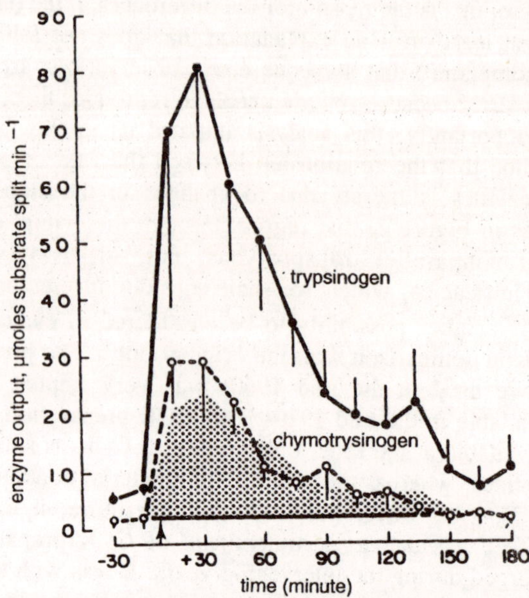

Figure 15.7 : Trypsinogen and chymotrypsinogen secretion from rabbit pancreas in organ culture in response to the COOH-terminal octapeptide fragment of CCK-PZ. Values are means ± SE for seven preparations.

these two enzymes. Thus, whatever others had found, we were able to reproduce our observation and this was gratifying.

But we did one additional experiment; one could say, one too many. It had been observed that many of the actions of CCK could be obtained with its carboxy-terminal octapeptide. The chemically synthesized "octapeptide" was available, and we tested it, full well expecting to see the nonparallel effect once again. Instead, we observed a parallel response in which the secretion of both enzymes was increased proportionately. That is, for a given trypsinogen secretion, the octapeptide produced a greater chymotrypsinogen response than CCK itself. Therefore, it appeared that CCK or a lingering contaminant in CCK preparations might have *inhibited* the chymotrypsinogen response. In recent studies by Niederau, Grendell and the author, we demonstrated that the presence of a CCK-receptor antagonist, proglumide, produces a dose-dependent selective inhibition of chymotrypsinogen secretion in the same biological system, suggesting that it is possible specifically to inhibit chymotrypsinogen secretion at the level of the CCK receptor. I shall discuss these experiments later on.

NONPARALLEL TRANSPORT

The parallel secretion model does not merely require that parallel secretion occur from time to time or circumstance to circumstance. It must always occur, in every circumstance that we measure, and for every possible condition, as long as transport, and not synthesis, determines the relative rate of secretion of the various proteins. Therefore, any single, clear case of nonparallel transport falsifies the hypothesis of *parallel transport* and requires any mechanistic model to be able to accommodate this fact. On the other hand, the view that transport is nonparallel does not require that it always be so. That would not make any sense, since the relative rates of transport of different proteins should always be parallel to each other in a given established state. It is only when state and circumstance change that nonparallel effects can come into play. That is, it is a "*permissive*" hypothesis in that whereas nonparallel transport may be seen in one situation or another, it. is not a necessary consequence of each and every circumstance, or even each and every change of state. Thus, one can provide all of the necessary evidence for the hypothesis that nonparallel transport exists and rejection of the hypothesis of *obligatory parallel* transport with a single confirmed example of *nonparallel transport*, such as the one that I have just discussed.

Nevertheless, it is useful to seek, even accumulate, additional cases of nonparallel transport, because however reliable our single observation may be, the reliability of our conclusion is greatly improved with each new independent discovery of nonparallel transport. Many such examples have been published since the mid 1960s. And unless we wish to consider them all measurement artifacts of one sort or another, the existence of nonparallel transport has beyond any reasonable doubt been affirmed. They represent about two dozen independent, statistically validated examples, each demonstrating nonparallel transport for different enzymes in different biological settings or demonstrating different aspects of the phenomenon. They include examples of nonparallel transport that are either directly or indirectly related to the concentration of digestive endproducts in the gut or blood, and therefore concern the regulation of specific digestive reactions in the intestine and are not solely "situational" examples of nonparallel transport, such as the CCK effect. My major reason for discussing this group of experiments is to explore in some detail different descriptive and analytic aspects of the phenomenon of nonparallel transport; that is, its intricacies. What exactly do we mean when we say that we have observed transport that is nonparallel in one circumstance or another? One needs to understand the nature of these phenomena in the most complete manner possible in order properly to evaluate and consider the ability of different mechanistic models for the secretory process to explain them. In addition, as I have already mentioned, knowledge of these details is also crucial in helping us choose the appropriate means of identifying nonparallel transport in the first instance.

A look at the published literature might lead us to conclude that it has become easier to find nonparallel transport than it used to be. Why should this be so? Even though the measurement techniques, enzyme assays and the like, have improved, the increased number of reports of such phenomena can be attributed solely, if at all, to the nature of the assays. If this is not the cause, then to what can we attribute the increased frequency of such observations?

It has been the common view that gastrointestinal functions are rather variable as biological activities go. The frequency of light to which a given photoreceptor responds is well defined and reproducible within relatively narrow limits from receptor to receptor and trial to trial. A particular type of muscle cell will shorten to approximately the same degree time after time, and even from muscle fiber to muscle fiber under the same load. However, if one examines a gastrointestinal activity and makes a measurement several times, particularly in situ, we would not be surprised to find rather different values from trial to trial. Although this may not be true for all alimentary canal functions, it is certainly the case for secretion, in particular for protein secretion. If we collected a dozen cases of spontaneous secretion in awake or even in anesthetised animals, the rate of protein secretion might vary among the animals or even within an animal from trial to trial and period of collection to period of collection by as much as an order of magnitude. This even applies to animals whose past history and present state are the same, as best as we can know it. This is of course another reason that it has become common practice to fast animals for defined periods before studying secretion, and to study the process at roughly the same time of day, etc., as well as to study secretion in vitro. Despite attempts at standardisation and however careful we may be, there nevertheless always remains a substantial variability in such processes in situ that we cannot avoid, or get rid of, because we do not know to what the variability is attributable.

The traditional method for determining whether or not nonparallel secretion has occurred is essentially the method that I presented above for the effect of CCK on protein secretion from the rabbit pancreas in vitro. The investigator averages output values for two or more enzymes at a particular period of time, that is, for a particular collection period, after a change in state and compares these values to control rates of secretion for the same enzymes. The question that is then posed is whether or not changes in the average rate of secretion of each enzyme subsequent to the imposed change of state have been equivalent or proportionate. If the changes in secretory rate were not proportionate, on the average, for different enzymes, then nonparallel transport occurred. If the average values changed in proportion for different enzymes, then transport was parallel. This method, although sometimes effective, is very insensitive because it does not take into account the large variability from trial to trial and animal to animal. Instead it incorporates this variability into the calculated variance around the mean values; values that are then compared with each other statistically. Hence a large and overlapping variance between populations whose means may differ substantially is often seen.

Many, and perhaps most, cases of nonparallel transport are likely to be missed if this method of analysis alone is used. This has commonly been the case. The reason for this "*insensitivity*" can be easily understood. The line drawn in this figure describes the relationship between the secretion of two different enzymes. The points closest to the origin would be for animals, trials, or periods of collection when secretion is low and those furthest away for animals, trials, or periods of collection when secretion rates are high. This variability from animal to animal and time to time is, as I have said, incorporated into the variance or standard deviation around the mean or average value for each period and over time, and in a given situation can be said to reflect variability in secretory activity of unknown origin from trial to trial. One can look at the figure and consider the array of points and estimate an average rate for each enzyme, relative to the range of values on the graph. We can now consider in our mind another line of twice or half the slope, over approximately the same range of values in which case one of the two enzymes will be present at twice the concentration relative to the other. Since we have made the slope twice ourselves, we can be confident that this is indeed true. However, if we seek evidence for this known fact by averaging the values for the enzymes under two different conditions and comparing such averages statistically, we can see that although the average value would change, the mean values would have large standard deviations, and we might well not be able to "see" the effect in statistical terms despite the known change. The large and overlapping variance would mask the true difference between the two conditions. Thus, real differences can be lost in our attempt at statistical validation by parametric averaging techniques. For this reason such analysis is weak in this situation, and many nonparallel events, particularly if they are subtle, that is, involve changes of two- or threefold or even more (if one can call that subtle) may well be missed.

We can eliminate this source of variance completely, without knowing its origin, merely by comparing least squares or regressed functions calculated from plots of data points for one condition relative to another rather than by comparing averages. This takes the variability from trial to trial into account and eliminates this source of variance from our statistical analysis, which is now a comparison between the parameters of two functions. In this way, our question can now be, "What is the ratio of one enzyme relative to another in secretion *for a given level of secretory activity?"* This simple device opens up, as we shall see, an abundant world of nonparallel transport events, that would without question never otherwise be noticed.

One can also improve upon averaging techniques by comparing average ratios of enzyme activity for any given trial, and in this way attempt to eliminate the variance in overall secretory activity by comparing only relative values. This is effective if, all of the points fit the line perfectly; that is, if there is no variance in the calculated function and hence no variation in the ratio from trial to trial. However, if substantial variation exists, then true differences may be missed if average ratios are merely compared parametrically. This would of course be a particular problem close to the origin, for example, for unstimulated secretion, where the two functions would converge. Once again, the use of correlation or regression analysis improves our ability to make such distinctions.

END-PRODUCT OF DIGESTION

The first time that this type of analysis was used was in studies on the effect of the amino acid lysine, added either to the intestinal lumen, blood, or extracellular fluid, on the relative rates of secretion of two proteolytic enzymes by the pancreas; one, *trypsinogen* whose active form splits proteins at lysine-containing peptide bonds, and the other, *chymotrypsinogen*, that does not.

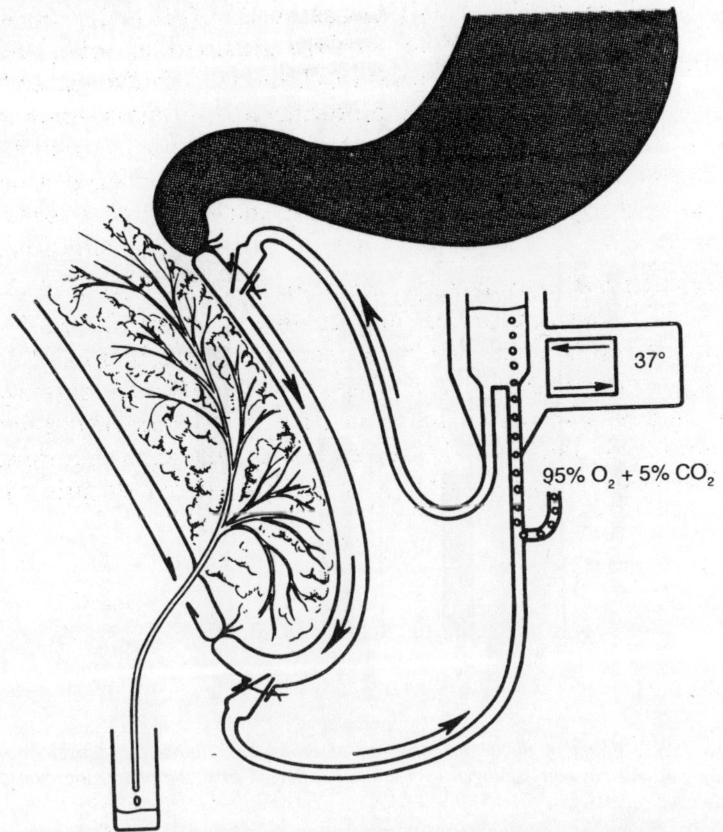

Figure 15.8 : Rabbit mesomental loop and pancreas indicating perfused portion of duodenum and cannulated pancreatic duct. Total volume of circulating fluid was between 75 and 100 ml.

However, briefly, when lysine was perfused through the duodenum of anesthetised rabbits at an effective concentration, the slope of the function relating the secretion of trypsinogen to chymotrypsinogen almost doubled; lysine had increased the secretion of trypsinogen relative to chymotrypsinogen, as compared to parallel controls. Similarly, in another series of experiments in which lysine was injected intravenously as a bolus, the slope of the function relating the secretion of the two enzymes was again increased, favoring trypsinogen secretion, this time by more than threefold. Lysine was also added to the medium bathing the rabbit pancreas in short-term organ culture, although at substantially higher concentrations. Once again, the relative rates of secretion of the two enzymes were altered, but in this case in the

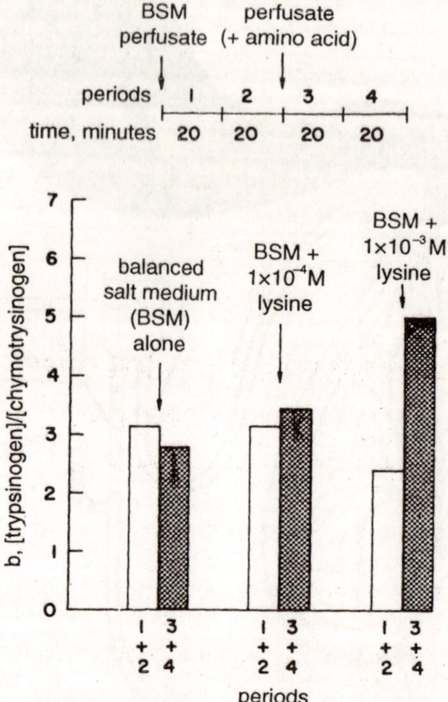

Figure 15.9 : Effect of duodenal perfusion of lysine on relative concentrations of trypsinogen and chymotrypsinogen in secretion collected from the pancreatic duct of anesthetized rabbits.

opposite direction, with chymotrypsinogen secretion being favored. It was also observed in these organ culture studies that in the control state, secretion became more trypsinogen dominant over time.

Thus, the *in situ* effect of the end-product lysine was a positive feedback response favoring trypsinogen secretion, whereas in vitro it produced a negative feedback response, the end-product increasing chymotrypsinogen secretion relative to *trypsinogen secretion*. This difference, however, may not indicate a different effector modality for lysine in situ and in vitro but merely differences in the concentrations of the amino acid used. Recently, Grendell demonstrated that lysine can cause the preferential release of trypsinogen (relative to amylase) from zymogen granules. The effect is one of positive feedback for lower concentrations of lysine, as in the in situ studies. However, when the concentration of lysine is increased to the levels used in the in vitro organ culture study, a negative feedback is observed, in which case the effect of the amino acid on trypsinogen release is lost. Thus, a concentrationdependent switch from positive to negative feedback occurs.

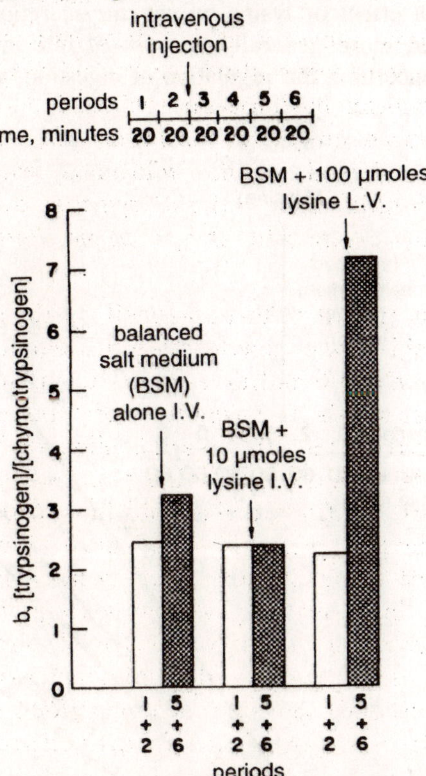

Figure 15.10 : Effect of intravenous lysine on relative concentrations of trypsinogen and chymotrypsinogen in secretion collected from the pancreatic duct of the rabbit in situ.

Beyond their regulatory implications, these observations on the effects of lysine demonstrated the usefulness of regression or correlation analysis in helping to discern nonparallel transport effects. In three separate series of experiments (gut perfusion, intravenous injection, and organ culture), lysine had produced nonparallel transport of the two enzymes that were measured. These effects could not have been distinguished statistically merely by comparing average period-by-period values for secretion, or even by comparing average ratios, because of the variability in secretion from animal to animal and tissue to tissue.

TRANSPORT OF CHYMODENIN

In the early 1970s Adelson purified a peptide taken from a side fraction of the CCK purification from hog duodenum. In light of the studies on the effect of lysine on enzyme secretion that I have just discussed, and more generally because of our interest in Pavlov's hypothesis concerning the regulation of digestion, Adelson set out to

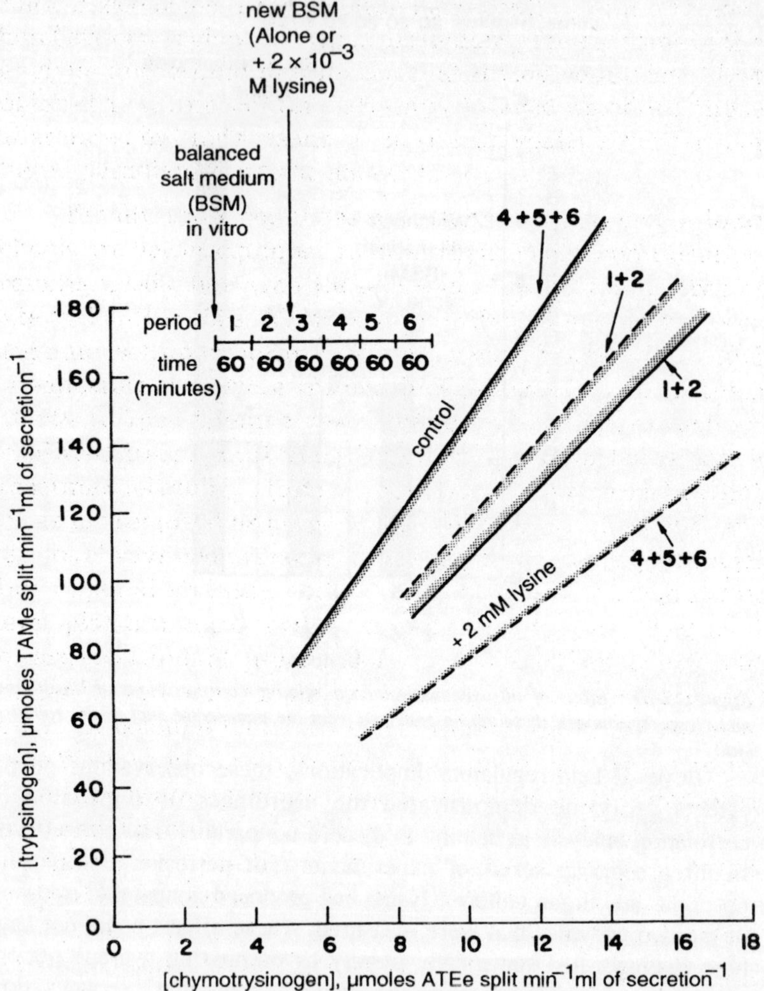

Figure 15.11 : Effect of lysine on relative concentrations of trypsinogen and chymotrypsinogen in secretion collected from the duct of the rabbit pancreas in organ culture.

determine whether or not the duodenal peptide that he had purified had a selective effect on the secretion of one or another digestive enzyme. Could it be a duodenal hormone that produces the selective secretion of one or more enzymes? He had already shown the peptide to be active in terms of pancreatic function, producing a modest overall enhancement of protein secretion in situ and the release of amylase from isolated zymogen granules.

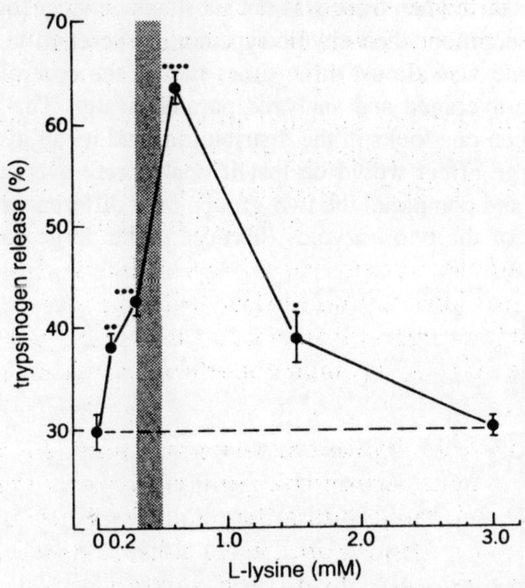

Figure 15.12 : Percent release of trypsinogen into a postmicrosomal supernatant of homogenised rat pancreatic tissue as related to the concentration of L-lysine in the homogenization medium.

After a preliminary assessment, we settled on two enzymes for further analysis, seemingly the least and most affected by administration of the peptide *in situ*, lipase and *chymotrypsinogen* respectively.

The results of the first series of experiments, carried out in rabbits *in situ*, were disappointing at first glance. The peptide did not appear to have an effect on the relative rates of secretion of the two enzymes, and the peptide-treated data points fell within the range of control values when individual measurements on the pair of enzymes were plotted one against the other. There was considerable variance in the proportions of the two enzymes from sample to sample in the control state, even though their relative rates of secretion were clearly and significantly correlated, and values for the peptide-treated samples were not distinguishable from control values comparing average outputs parametrically. However, a nonparametric rank sum test showed that peptide-treated points had a higher chymotrypsinogen output than control points. A closer look at the data revealed that the peptide-treated points were not randomly distributed among the broad range of control values but tended to form a border of the distribution along the chymotrypsinogen axis. We found that if we only considered samples of secretion for which lipase output was relatively low, that is, samples taken when there was not substantial endogenous stimulation of protein secretion, then chymotrypsinogen secretion in the presence of the peptide was almost three times that in the control group when values were averaged and analyzed parametrically. This difference is apparent when one looks at the distribution, and it can also be realised that this large effect would be lost if one merely averaged all of the data points and compared the two groups or if differences were sought in the ratio of the two enzymes (because of the large variance in the control state).

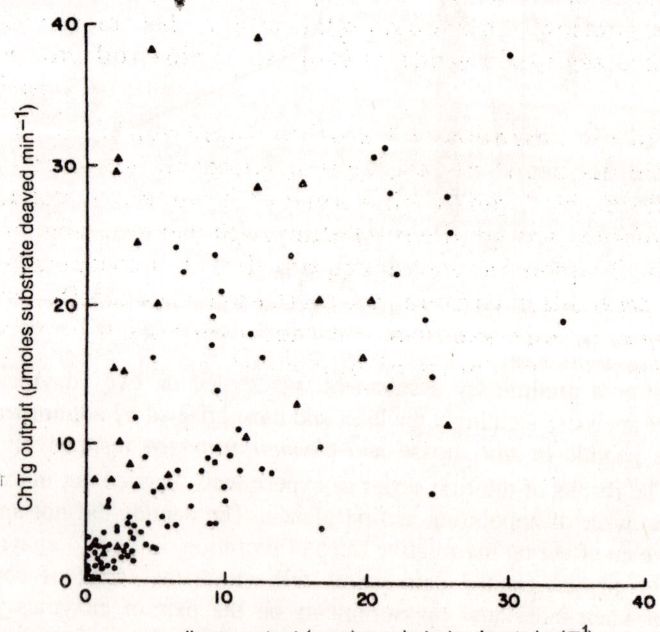

Figure 15.13 : Total distribution of data points for individual collection periods displaying concomitant ChTg (chymotrypsinogen) and lipase outputs into the duct from rabbit pancreas in situ. Axes give micromoles substrate cleaved per minute per entire sample collected in 20 min. (A) Chymodenin treated; (•) saline-treated controls: nine animals in each group.

We repeated the study in vitro using the organ preparation of rabbit pancreas and observed essentially the identical effect. Once again the peptide-treated points formed a border along the chymotrypsinogen axis of a loosely correlated control distribution. However, in this case, it was possible to demonstrate changes in the average rate of chymotrypsinogen secretion in the absence of changes in lipase secretion, using a period by period analysis of output values. Relative to parallel controls, chymotrypsinogen secretion had been increased by some fourfold, whereas lipase secretion was unaffected by the addition of the peptide. Thus, the peptide favored the secretion of chymotrypsinogen relative to lipase and for this reason was named chymodenin; chymo- for chymotrypsinogen, and -denin for its tissue of origin, the duodenum.

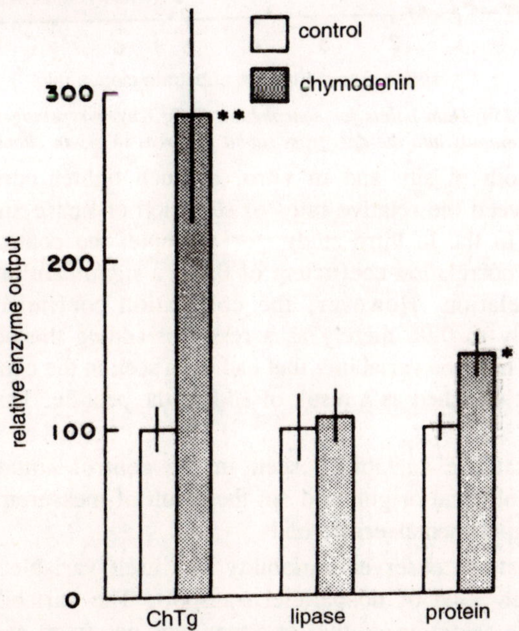

Figure 15.14 : Effect of chymodenin on output of chymotrypsinogen (ChTg), lipase, and total protein into the duct from rabbit pancreas in situ. Mean output ± SE of mean are given.

There was another aspect of the effect that was quite interesting, as well as unexpected. The control data displayed considerable variability in terms of the relative rates of secretion of the two enzymes, and the amount of one enzyme secreted was not a good indicator of the rate of secretion of the other. However, after the addition of the

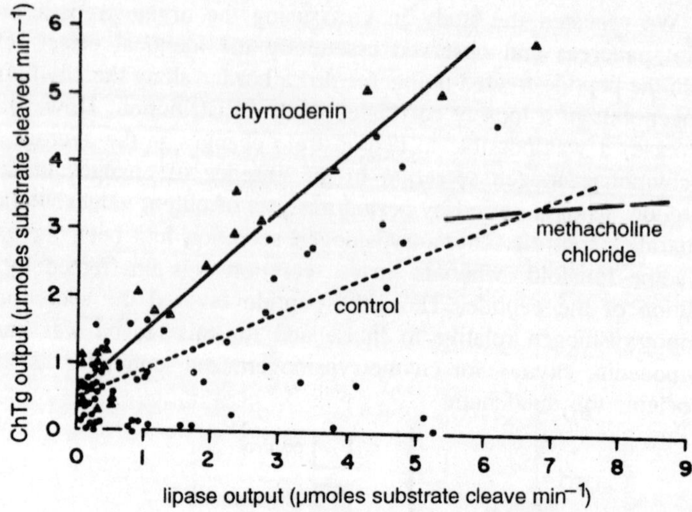

Figure 15.15 : Data points for concomitant ChTg (chymotrypsinogen) and lipase outputs into the duct from rabbit pancreas in organ culture.

peptide, both in situ and in vitro, a much tighter correlation was found between the relative rates of secretion of lipase and chymotrypsinogen. In the in vitro study, for example, the control regression line had a correlation coefficient of 0.53; a significant but not a very tight correlation. However, the correlation coefficient increased dramatically to 0.96 merely as a result of adding the peptide to the medium. Thus, the variability that had been seen in the control situation had almost vanished as a result of adding the peptide. This meant two things:

(1) that the variability seen in the control situation was of biological origin, and not the result of measurement or other experimental error; and

(2) that the observed variability was itself variable. This was a new kind of nonparallel transport. The variability between the secretion of two enzymes was not fixed as the parallel transport hypothesis required, but rather the extent to which the amount of one enzyme secreted could predict the amount of another being secreted at the same time was itself variable.

Thus, we had found two additional cases of nonparallel transport in the effect of chymodenin on the relative rates of chymotrypsinogen and lipase secretion in vitro and in situ. In addition, we found that the degree to which the secretion of different enzymes covary is also

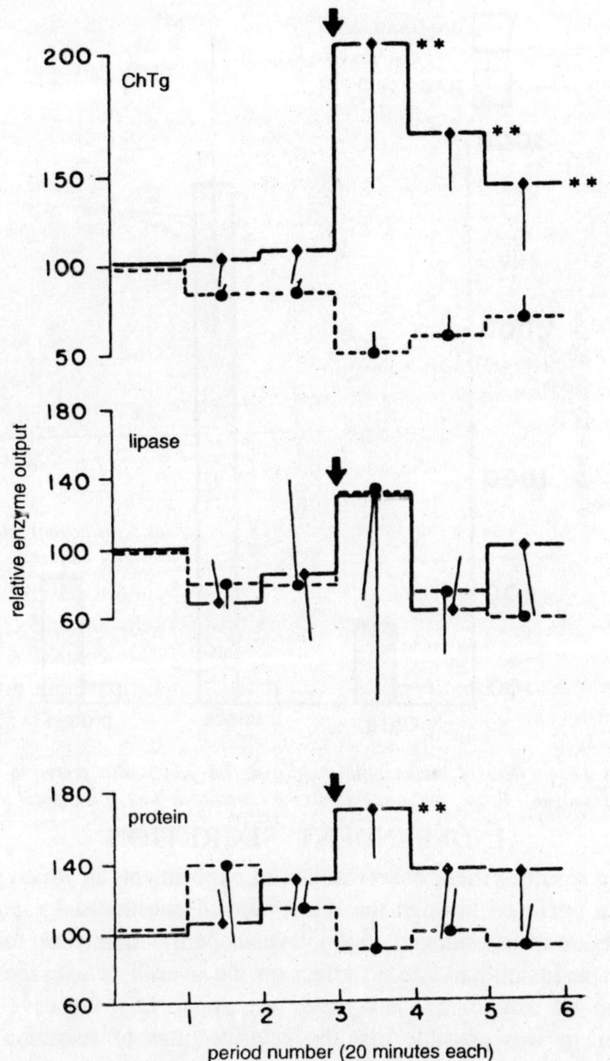

Figure 15.16 : Effect of chymodenin on ChTg (chymotrypsinogen), lipase and protein output into the duct from rabbit pancreas inorgan culture.

variable. There was an additional nonparallel transport effect observed en passant, if you will. When a cholinergic agonist was added to the medium bathing the pancreas in organ culture, there was a nonparallel response in which lipase secretion was favored relative to chymotrypsinogen, unlike the effect of chymodenin.

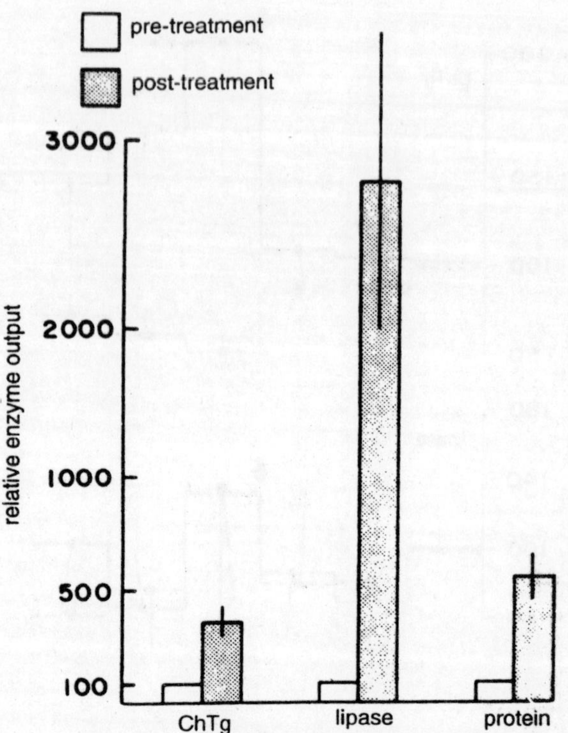

Figure 15.17 : Effect of methacholine chloride on pancreatic secretion of ChTg (chymotrypsinogen), lipase, and protein into the cannulated duct of the gland in culture.

INDEPENDENT SECRETION

As a result of these observations, of experiments in which glucose had been perfused through the duodenum of anesthetised rabbits and the secretion of trypsinogen, chymotrypsinogen, and amylase followed. Glucose perfusion had had no effect on the overall or average rate of secretion of any of the enzymes, an apparently negative study. However, it was possible that the *relative* rates of secretion of the different proteins might have been altered. When a regression analysis was performed such relative changes could be demonstrated for each of the three enzyme pairs. In each case the calculated slope had been altered by about two to three times as a result of glucose perfusion. Therefore, we had three more cases of nonparallel transport, this time involving three enzymes in one situation, the secretion of each varying relative to the others.

Table 15.3 : Effect of Duodenal Perfusion of 5 mM Glucose on Proportions of Three Enzymes in Secretion as Determined by Linear Regression of Enzyme Pairs

	Slope, *b*					Intercept, *a*	
	-Glucose	r	+Glucose	r	Δ%b	-Glucose	+Glucose
Trypsinogen/	1.8	0.89	5.6	0.83	208%	5.1	0.3
chymotrypsinogen	±0.3		±1.2			±1.3	±17
Chymotrypsinogen/	2.0	0.86	0.9	0.96	117%	0.6	0.6
amylase	±0.4		±0.1			±0.7	±0.1
Trypsinogen/	3.4	0.73	6.5	0.91	91%	6.3	6.5
amylase	±1.0		±1.0			±1.8	±1.0

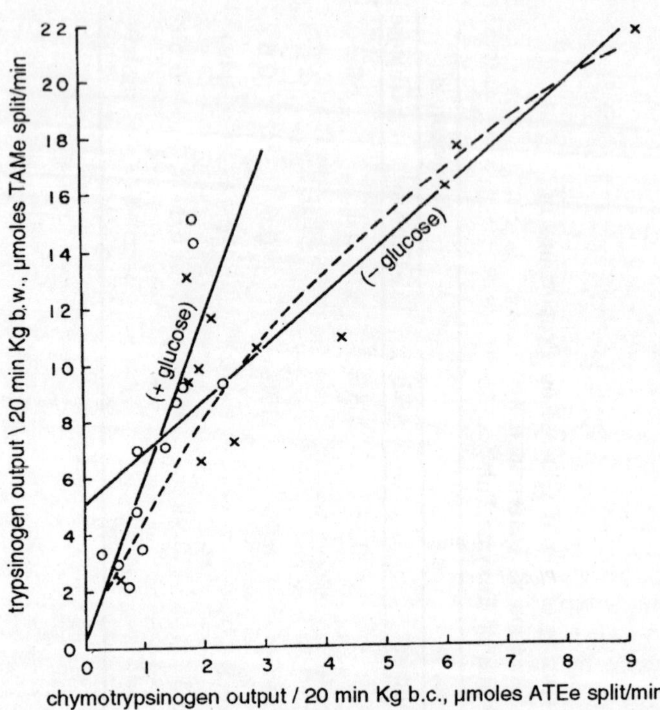

Figure 15.18 : Plot of secretion of enzyme pair chymotrypsinogen-trypsinogen into the cannulated duct of the rabbit pancreas in situ before glucose was added to perfusion medium (X)

Yet another type of nonparallel transport became evident from these experiments. In several cases the calculated regression lines did not go through the origin but intercepted the axes at points significantly different from 0,0. This could mean one of two things. The secretion of one enzyme could occur in the complete absence of the secretion of the other. This would, of course, be evidence that the secretion of different enzymes occurred independently and was not linked obligatorily in some fashion, as in the exocytosis of granules of mixed enzyme content. Alternatively, the function describing their relative rates of secretion might be nonlinear and the fact that the line did not regress through the origin might therefore simply reflect the presence of a curvilinear function. In this case, the relative rates of secretion would vary continuously (be continuously nonparallel) as a function of the overall rate of secretion. Such behaviour had been documented in other cases, one of which I discussed above, and others of which I will consider below. In this case, it was not possible to distinguish

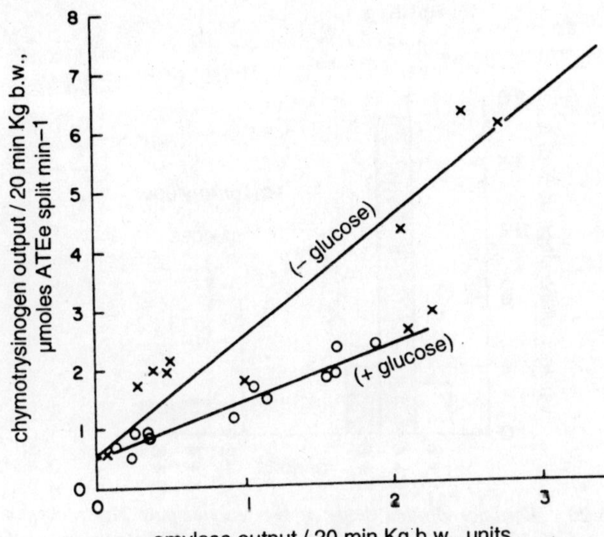

Figure 15.19 : Plot of secretion of the enzyme pair amylase-chymotrypsinogen into the cannulated duct of the rabbit pancreas in situ before glucose was added to the perfusion medium (X)

between the two possibilities. Of course, both effects might occur, nonlinear functions and positive intercepts, either separately or in the same function. Although it could not be determined whether the control function in Figure was either a straight line or a curve, for the "+ glucose" line in Figure elsewhere in this chapter, in which case the data points approach the axis closely, a true positive y-axis intercept seems clear, indicating that the secretion of chymotrypsinogen can occur independently of amylase.

In addition, it was found that glucose perfusion could alter the apparent intercept as well as the variance around it. Glucose perfusion eliminated the calculated y-axis intercept for the trypsinogen-chymotrypsinogen pair, and the data now regressed through the origin. For amylase-chymotrypsinogen, although glucose perfusion did not alter the numerica. value of the y-axis intercept, it reduced the variance around the observec intercept so that it became statistically distinguishable from the origin unlike the control state. This reflects the fact that glucose perfusion had increased the covariance of the secretion of the two enzymes, another example of a treatment that further entrains the secretion of twc enzymes.

But perhaps the most interesting effect seen in this study was

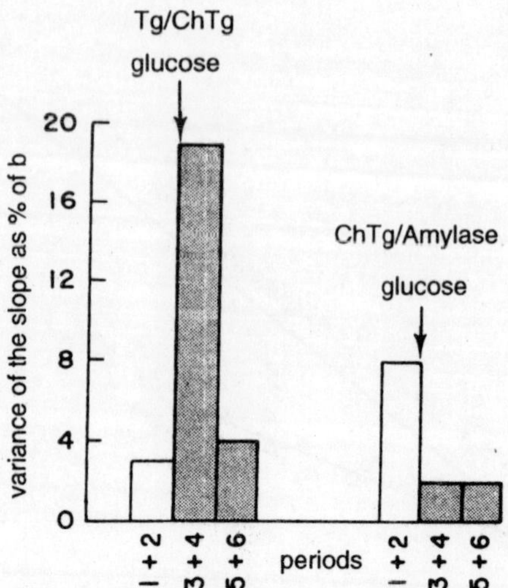

Figure 15.20 : Variance around slope of two enzyme-pair plots, chymotrypsinog-entrypsinogen (Tg/ChTg) and amylasechymotrypsinogen (ChTg/amylase), before (clear bars) and after (shaded bars) addition of glucose to perfusion medium.

related tc changes in the degree of covariance in the secretion of *three enzymes.* If the extent of covariance in the secretion of different enzymes was due tc variability in the secretion of enzyme from granule pools of different mixec enzyme content, then changes in the variability of selection among these different granules, either increases or decreases, would occur in parallel foi all enzymes. Only if the proteins were secreted independently of each other would it be possible, for example, for there to be a change in the covariance of chymotrypsinogen secretion relative to trypsinogen in one directior (more or less variable), and a change for chymotrypsinogen secretior relative to amylase in the opposite direction. This is exactly what wa, observed. For the first two periods after the initiation of glucose perfusion there was a large decrease in the entrainment of the secretion of trypsinoger and chymotrypsinogen, that is, they were secreted more independently of each other. At the same time entrainment in the secretion of chymotrypsinogen and amylase became much greater. In the former case, the variance increased some fivefold, whereas in the latter it was decreased by about 75%. These observations presented compelling evidence for the hypothesis that the proteins were secreted independently of each other, whatever the mechanism of secretion might be.

FLUCTUATION

Several years ago Tseng, Grendell, performed a series of experiments in which we attempted to repeat earlier work by a Swiss group who had demonstrated that extracts of rat duodenum, taken from animals previously fed either a glucose meal or a protein hydrolysate "*lactalbumin*", injected directly into the intrapancreatic circulation via the celiac artery led to the enzymespecific secretion of amylase or trypsinogen, respectively. They proposed the existence of enzyme-specific hormones in the duodenal mucosa, a class of "*chymodenins,*" if you will.

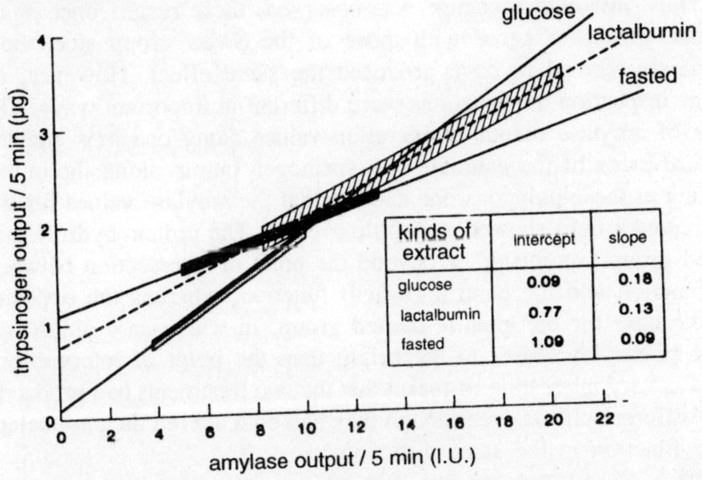

kinds of extract	intercept	slope
glucose	0.09	0.18
lactalbumin	0.77	0.13
fasted	1.09	0.09

Figure 15.21 : Effect of intraceliac injection of duodenal extracts from fasted, lactalbumin-fed, and glucose-fed rats on relative proportions of amylase and trypsinogen found in pancreatic secretion collected from the cannulated duct of anesthetised rats.

We were unsuccessful in reproducing exactly what they observed, but were able to find a similar, although less dramatic, phenomenon that was consistent with their observation and whose characteristics were enlightening. Unlike the Swiss group's report, in our hands the injection of extracts from glucose-fed rats did not increase the secretion of amylase when periodby-period averages were compared. And although protein hydrolysatetreated extracts did increase trypsinogen secretion by about twofold, it produced a roughly parallel increase in the average rate of amylase secretion as well. We pursued the matter further by regression analysis. When the secretion of amylase was plotted against that of trypsinogen for individual samples, we found that both treatments had altered the regression lines. The function for extracts taken from fasted rats (controls) had a substantial positive y-axis intercept (trypsinogen secretion in the absence of amylase). The regression line calculated for data from rats injected with the glucose-exposed extract was altered in two ways. The slope of the function was about doubled, favoring trypsinogen secretion, and the intercept became indistinguishable from one passing through the origin. The effect of the extract from protein hydrolysate-fed rats was similar in that it also both lowered the y-axis intercept toward the origin and increased the slope of the function, although it did both to a lesser degree.

Thus, although a change was observed, these results once again did not appear to agree with those of the Swiss group since both treatments seemed to have produced the same effect. However, on further inspection the functions were different in important ways. The range of amylase output or secretion values along one axis and the standard error of the estimate of trypsinogen output along the other. Looking at these polygons one can see that the amylase values for the two treated groups showed very little overlap. The protein hydrolysate-treated group lay outside or beyond the point of intersection between this function and the control (fasted) function, whereas the opposite was the case for the glucose-treated group, in which case almost all of the points lay closer to the origin than the point of intersection. This was very interesting. It meant that the two treatments had produced .very different effects even though they had both altered the parameters of the function in the same direction.

The control and experimental functions intersected each other. The reason for this is that treatment had altered both slope and intercept. In such a case, a single set of changes in the parameters of a function can produce different effects depending upon the location of the data points along the function; that is, depending upon the overall rate of secretion. For points closer to the origin than the intersection of the control and experimental lines, the intercept effect dominates and, in our case, that favors amylase secretion; whereas for points further away from the origin than the intersection, the slope effect dominates and trypsinogen secretion would be favored. Interestingly, the data from the protein hydrolysatetreated group lay primarily outside of the point of intersection, favoring trypsinogen secretion, whereas data from the glucose-treated group lay closer to the origin, favoring amylase secretion. Thus, in the end the data did support the observations of the Swiss group.

But beyond this they also demonstrated a curious aspect of the phenomenon of nonparallel transport that is important to appreciate.

If a treatment changes a function so that the new function intersects the old, this single change in a function's parameters can lead to an increase in the proportion of one enzyme in secretion, relative to another, for values outside the point of intersection, produce no change at all if the data points fall in the area of the intersection between the two lines, or produce a decrease for values closer to the origin than the intersection. Therefore, a single treatment effect can alter the relative proportions of the two enzymes in secretion in one direction, in the opposite direction, or not at all, depending upon how active secretion is overall.

ENERGETICS

In the same study, CCK was injected into anesthetised rats and the secretory response for amylase and trypsinogen measured. A dramatic nonparallel effect was produced in which the ratio of amylase to trypsinogen increased by about three times. A similar nonparallel effect had been reported earlier by others for a different enzyme pair in response to CCK in the rat. When the individual data points

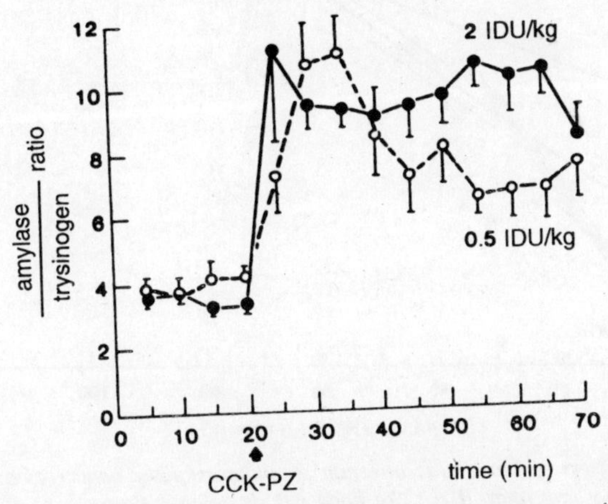

Figure 15.22 : Ratio of amylase to trypsinogen in pancreatic secretion collected from the cannulated duct of anesthetised rats before and after intraceliac injection of cholecystokininpancreozymin.

were plotted, amylase versus trypsinogen, it became clear that the function relating the secretion of the two enzymes was not linear, but became more and more amylase dominant as overall enzyme secretion became greater. That is, the proportions of the two enzymes varied continuously as a function of the overall output of enzyme in what was presumably the same state (after the injection of CCK).

This was not the first time that we had seen such a relationship. We had reported the existence of a nonlinear relationship between the secretion of two enzymes in 1974, for the effect of cholinergic stimulation on the secretion of trypsinogen and chymotrypsinogen by the rabbit pancreas in situ, as well as for the original nonparallel response to CCK, which also displayed the same type of nonlinear relationship. The former (cholinergic) function became more chymotrypsinogen dominant as overall output became greater (the ratio of trypsinogen to chymotrypsinogen being 2.5 ± 0.3 SE for the highest one-third of observed values, and 5.7 ± 0.7 for the lowest third), whereas the latter (CCK) function became more trypsinogen dominant as overall output became greater.

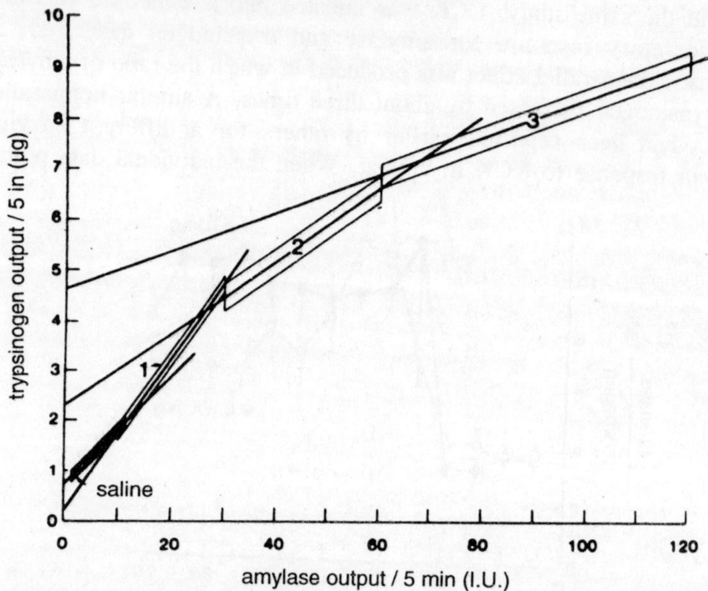

Figure 15.23 : Effect of intraceliac injection of cholecystokinin-pancreozymin [CCK-PZ, 0.5 or 2.0 Ivy dog units (IDU) /kg body wt] on relative proportions of amylase and trypsinogen in pancreatic secretion collected from the cannulated duct of anesthetised rats as compared with values for saline-injected controls.

Before we move on, one additional point about the effect of CCK on *amylase* and *trypsinogen* secretion by the rat pancreas is worth noting. When control and experimental lines intersect, then a single treatment may favor the secretion of one enzyme closer to the origin than the point of intersection of the two lines, may favor neither in the region of the intersection, and may favor the other enzyme beyond it. Thus, in this case there are three different potential effects of one treatment, depending upon the overall level of secretory activity. The situation becomes even more complex if the function is a curve. In this case, as the data points are farther and farther removed from the origin beyond the point of intersection of the two functions, the ratio of enzymes again begins to approximate more and more closely the control function as the curve falls off. Indeed, the slope may become less than that of the control function. This effect is produced by CCK in the rat. At the intersection, CCK produces no change at all. Beyond the intersection of control and CCK functions trypsinogen secretion is favored (increased slope) and then as the function falls off at higher rates, amylase secretion becomes increasingly favored so that control ratios may again be observed or even reversed toward a more amylase-dominant slope. Thus, moving out from the origin, one treatment, CCK, produces five distinct, continuously varying effects on the ratio of the two enzymes relative to the control, depending upon the overall rate of secretion.

Although a good fit to a linear function is frequently obtained for plots of enzyme pairs close to the origin, it has become increasingly clear that a single general nonlinear function may describe the relationship between the rates of secretion of any enzyme pair for all rates of secretion. The best fit has thus far invariably been to a simple power function of the form, y = axb. This may be significant because this simple relationship connecting the rates of secretion of any two enzymes is consistent with a model for secretion in which the relative rates of secretion of the individual enzymes are independently determined as a function of concentration gradient and permeability; the parameters of the function merely reflect the different kinetic parameters at work for the different enzymes.

In two recent studies, Tseng, Grendell, have found numerous examples of nonparallel transport in which the parameters of such power functions are altered by various physiological stimuli. For example, in the presence of CCK the injection of an effective dose of either glucose or lysine into the celiac artery of the rat leads to the selective secretion of trypsinogen or amylase, respectively. When these data are plotted and the rates of secretion of the two enzymes compared, the CCK curve is shifted toward the amylase axis as a result of glucose injection and toward the trypsinogen axis as a result of lysine injection.

In the other series of experiments, secretion of the same two

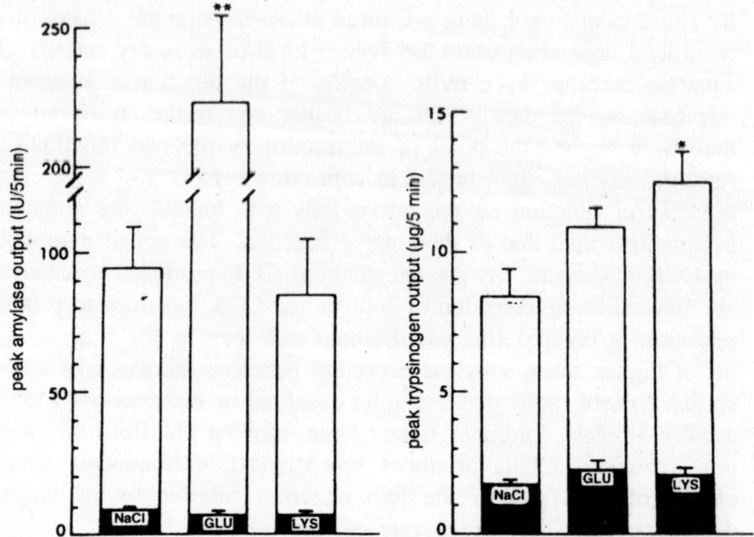

Figure 15.24 : Peak amylase and trypsinogen output from pancreaticobiliary duct of anesthetized rats following injection into the celiac artery of 250 μl of 0.9% NaCl (NaCl), NaCl containing 15 MM D-glucose (Glu), NaCl containing 0.5 MM L-lysine (Lys), or NaCl containing 0.5 IDU/kg body wt cholecystokinin (CCK). CCK + Glu or CCK + Lys indicate combinations in NaCl at doses specified.

enzymes was followed in the presence of either insulin or glucagon, along with CCK.

Both hormones increased the rate of trypsinogen secretion relative to that of amylase. However, they did so in opposite ways. Insulin increased trypsinogen output, whereas glucagon decreased amylase secretion. When the data points were plotted, the curves were again shifted; this time both treatments produced more trypsinogen dominant functions in a dose-dependent manner. However, points along the insulin curve for the most effective dose tended to lie further from the origin than data points from the glucagon experiments, which tended to lie closer to the origin. This produced the two different treatment responses, insulin increasing the amount of trypsinogen secreted and glucagon decreasing amylase secretion. In addition, as the dose of either insulin or glucagon was increased beyond that most effective, the effect was diminished and the values and functions returned to or toward the control. In these experiments each and every separate curve represents two types of nonparallel transport: the first, in its own right merely by dint of being a curve, and the second, relative to the parameters of the control (CCK alone) function.

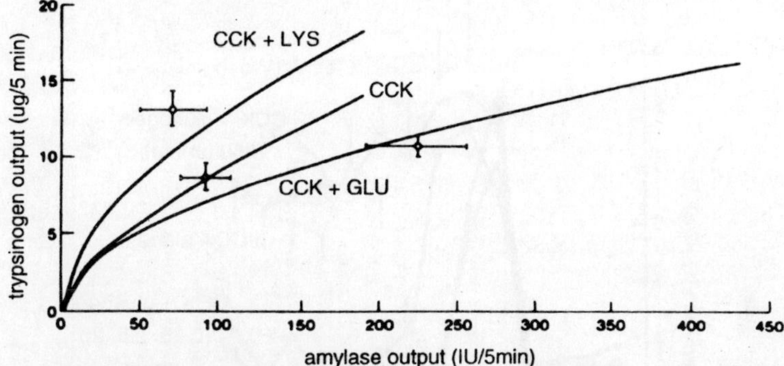

Figure 15.25 : Regression lines calculated from scatter of individual pairs of enzyme measurements for all data points over a 50-min period following injection of CCK, CCK + Glu, or CCK + Lys.

Nonparallel transport was demonstrated pursuant to the inhibition of secretion, in particular, inhibition of CCK-stimulated secretion by the rabbit pancreas in organ culture by the glutaramic acid analog, proglumide. These experiments demonstrate two interesting effects.

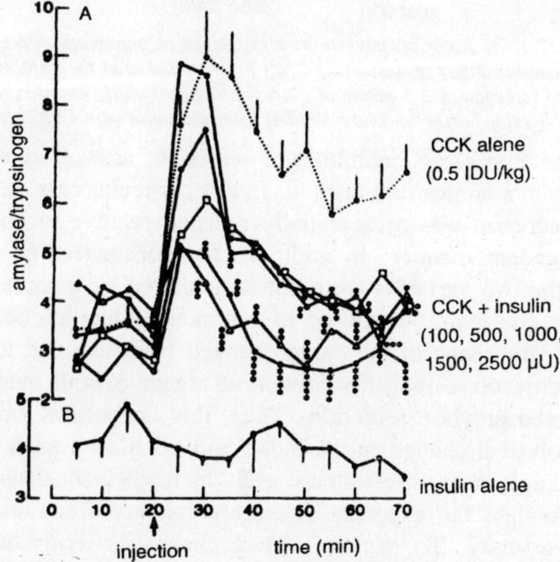

Figure 15.26 : (A) Ratio of amylase to trypsinogen in pancreatic secretion collected from the cannulated duct of anesthetised rats prior to and after intraceliac injection. (B) Effect of injection of 500 μU of insulin alone. Points, means; error bars, SE. Because of overlap, error bars are omitted from all points for CCK + insulin groups before and after injection.

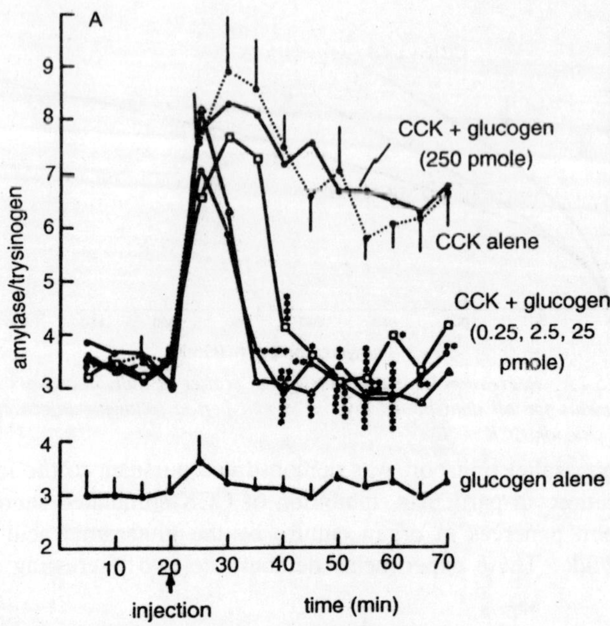

Figure 15.27 : (A) Ratio of amylase to trypsinogen in pancreatic secretion collected from the cannulated duct of anesthetised rats prior to and after the intraceliac injection. (B) Effect of injection of 2.5 pmole of glucagon alone. Points, means, error bars, SE. Because of overlap, error bars are omitted from glucagon with CCK groups.

In the first place, inhibition of secretion, as well as stimulation, can occur in a nonparallel fashion. In this particular case, chymotrypsinogen secretion was preferentially reduced relative to amylase in a dose-dependent manner. In addition, the parameters of the curves relating the two variables were not only altered by treatment, as with insulin or glucagon, but altered so that as the inhibition became more profound, the shape of the curve changed from one that increasingly favored chymotrypsinogen secretion at higher overall output, to one that favored amylase secretion. Thus, this nonparallel transport not only involved a change in the slope and pitch of a particular curve but a change in its direction as well. In mathematical terms this is simple enough; the exponent is merely changed from one above to one below unity. To explain such a change in terms of a vesicle mechanism requires complex models. On the other hand, in an equilibrium-based transport process, this effect would simply reflect a change in the kinetic parameters for the independent transport of the two proteins.

BORDERS OF THE PARADIGM

Two approaches have been taken to deal with observations of nonparallel transport from a paradigmatic perspective. The first is that they are merely experimental artifacts, and that accurate measurements of secretion would show it always to be parallel. Such claims of artifact are common in science and particularly so when an established idea is being questioned. However, beyond the claim, more than 15 years after our first report of nonparallel transport,

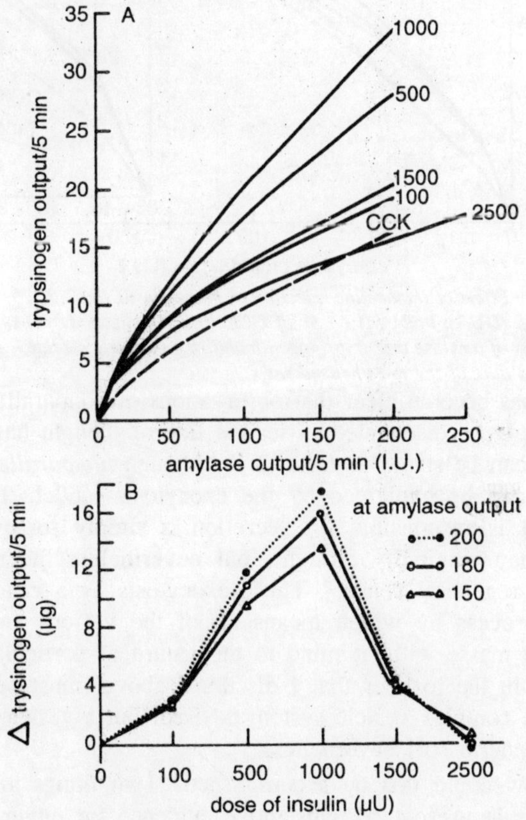

Figure 15.28 : Effect of intraceliac injection of insulin (100, 500, 1000, 1500, or 2500 µU) with CCK (0.5 IDU/kg body wt) (_____) or CCK (0.5 IDU/kg body wt) alone (-------) on relative proportions of amylase and trypsinogen found in pancreatic secretion collected from the cannulated duct of anesthetised rats.

during which time there have been many and varied reports of this phenomenon, there has not been a single example of a nonparallel transport observation that has been documented as being due to a particular artifactual cause. At this point it is hard to imagine that the range of observations that have been made.

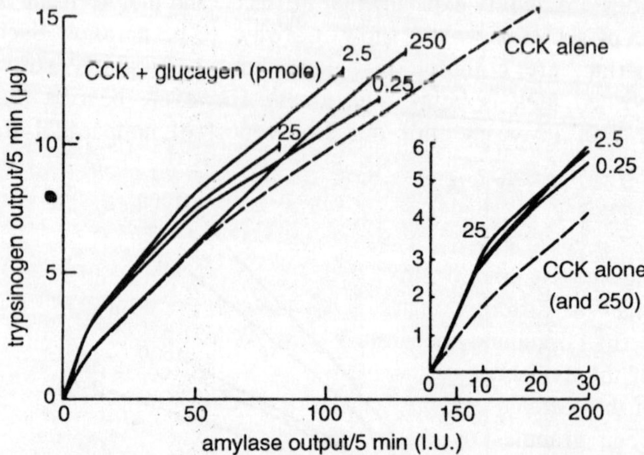

Figure 15.29 : Effect of intraceliac injection of glucagon (0.25, 2.5, 25, or 250 pmol) with CCK (0.5 IDU/kg body wt) (-) or of CCK (0.5 IDU/kg body wt) alone (——) on relative outputs of amylase and trypsinogen found in pancreatic secretion collected from the cannulated duct of the anesthetised rat.x

As it has become clear that observations of nonparallel transport cannot simply be discounted, a second line of thought has emerged. This view can be stated as follows: "Although *nonparallel transport* occurs, it can be explained by the exocytosis model. The vesicle system that is responsible for secretion is simply somewhat more complex than originally thought, but nevertheless the model still applies in a general sense." Thus, exocytosis as a random, mass transport process by which means all of the various enzymes are released en masse without mind to the nature of particular secreted molecules, in the manner that I discussed above, must be put aside and a more complex vesicle system devised, but a vesicle system of the same general sort nevertheless.

But how could this be accomplished? Two things are required and there is at present no substantive evidence for either. The first is that different granules would have to contain different enzymes either singly or in combination. The second is that the secretory mechanism would have to be able to distinguish or select between these different granules either within *single cells or between cells. There is no evidence other than the circular "evidence"* of nonparallel transport itself, to suggest that individual granules contain single

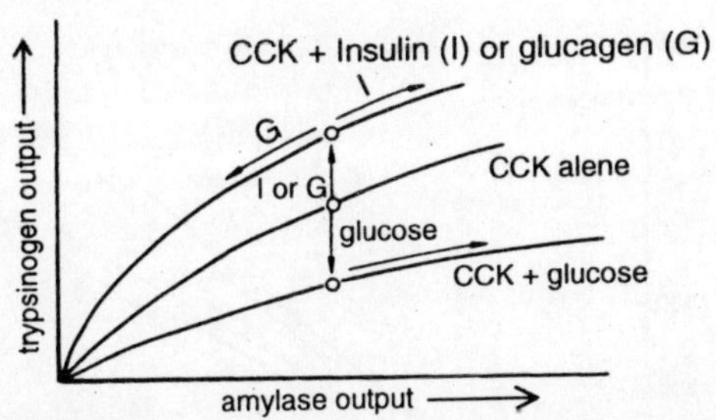

Figure 15.30 : Idealized schema showing effect of insulin (I), glucagon (G), and glucose on relative rates of amylase and trypsinogen secretion by rat pancreas in presence of CCK.

species of enzyme or that there are ordered differences between granules of different mixed enzyme composition. Indeed, from the very first immunocytochemical antibody studies to the most recent, all of the evidence suggests that each granule and each cell contains all of the secreted enzymes, and ordered differences in enzyme content between granules or cells have never been reported. Regional or geographic organisation has not been demonstrated, although small differences have been reported between the enzyme content of acinar cells surrounding the islets of Langerhans, about 1% of the gland mass, and the remaining 99%. Such differences probably cannot account for a measurable nonparallel effect, no less for the variety of such phenomena that have been reported, or for the varied characteristics of nonparallel transport that I have discussed.

In one reported case of *nonparallel transport*, the author proposed that the stimulant that had produced the effect elicited secretion of enzyme from a group of cells that contained one mixture of enzyme, whereas basal or unstimulated secretion was derived from another group of cells with a different mix of enzymes whose secretion was not augmented by the stimulant. Of course, it is important to remember that the existence of the nonparallel effect itself did not provide evidence for this hypothesis. This explanation was an hypothesis proposed in an attempt to account for the observation within the borders of the exocytosis model for secretion.

How easily can we explain such observations and those that I have discussed above in terms of a model of secretion based on ordered intercellular or intergranular differences in enzyme content? We can begin by considering the proposal in terms of the power function described above. For example, such a function could be due to the variable admixture of material from two groups of granules

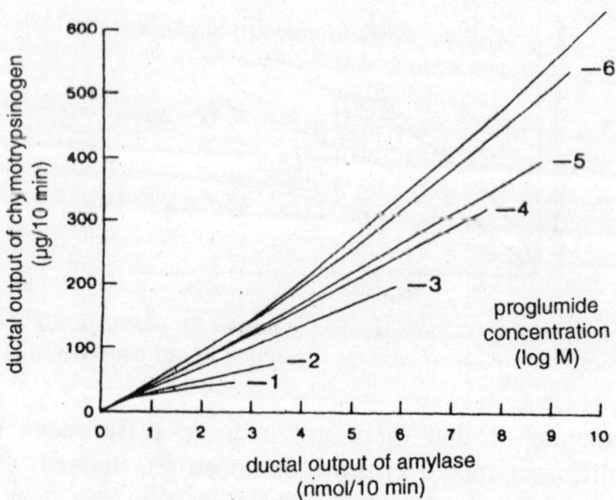

Figure 15.31 : Regression lines calculated from the scatter of all individual pairs of enzyme outputs into the cannulated duct of the rabbit pancreas in organ culture for each different dose of proglumide (in the presence of CCK).

or cells with different enzyme content. In such a situation, the steepest portion of the function, that closest to the origin, would most closely approximate the enzyme content of one pool and the portion of the curve farthest away from the origin would most closely reflect the contents of the other pool. From some of the figures presented in this chapter we can see that such source pools would have to differ from severalfold to as great as an order of magnitude in their content of different enzymes. Such differences would probably be obvious with immunocytochemistry, but have yet to be observed, or at least reported. But there are additional problems. It is possible to change the direction and parameters otherwise of such curves, and this requires further regulatory modulation as well as additional enzyme pools for each pair of enzymes. The central complexity of such a model is in the number of pools required to account for all of the potential possibilities. In order to account for the coordinated variable secretion of all of the enzymes in a continuous fashion in this manner, the number of different pools containing enzyme in different ordered proportions would be of the order of 19!-the number of secreted proteins minus one factorial. This is more than the number of cells or granules in the pancreas of a rodent. But even if we consider that the number of permutations is limited to only *pairs* of enzymes, the number of different pools would still have to be in the hundreds.

We must also consider this explanation in light of other phenomena such as positive intercepts which demonstrate that one enzyme can be secreted in the absence of another, and the observation that the extent of the covariance of the secretion of two particular enzymes may increase, while at the same time the covariance between one of these enzymes and a third species may decrease. That is, these changes do not suggest mixing from fixed pools of mixed enzyme content. One could go on, but of course it is always possible to imagine some manner of means, however complex, to explain such results in vesicular terms. On the other hand, what evidence there is on hand does not support the existence of either intergranular or intercellular organisation, and the sole reason for thinking that such organisation exists is the underlying belief that protein secretion must occur by exocytosis. If this theoretical "crutch" is removed, then certainly the burden of proof must be on the proposal for such a complex system.

If in all of the studies, only parallel transport had been seen, namely all lines regressed through the origin, no changes in intercepts or slopes of functions were observed, no changes in variance occurred, no curves were formed, and so forth, then the conclusion that parallel transport applied in general and that there was no reason to question the mass transport exocytosis model would have reasonably been drawn. The rejection of each prediction of the parallel transport model requires that the exocytosis model be substantially modified, namely made more complex, and that the negative evidence regarding the existence of ordered pools be put aside to await positive evidence, in order to save the vesicle model.

On the other hand, the many complexities of mechanism visited upon the vesicular model by nonparallel transport are simply, even trivially, explained by a model in which different enzymes move independently of each other as a result of their own kinetic energy. All of the observations can be accounted for as merely being due to changes in the transport parameters of concentration gradient and permeability. Beyond this we must remember that the mechanism underlying nonparallel transport must be considered in light of all the other evidence.

16

Enzymes of Protein Metabolism

Proteinase K is a serine protease and the main proteolytic enzyme produced by the fungus *Tritirachium album* Limber (1). The enzyme has a broad specificity, cleaving peptide bonds C-terminal to a number of amino acids. The enzyme is produced, together with other proteases and an aminopeptidase, during stationary phase when the fungus is grown by submerged culture. The enzyme is so named because the organism can grow on native keratin as sole carbohydrate and nitrogen source owing to the enzyme's ability to digest keratin. Because of its broad substrate specificity, high activity, and its ability to digest native proteins, proteinase K has found considerable use in procedures where the inactivation and degradation of proteins is required, particularly during the purification of nucleic acids. The enzyme retains activity in the presence of 0.5% sodium dodecyl sulfate, which is used in mammalian cell lysis. This allows the use of proteinase K in conjunction with cell lysis resulting in the rapid degradation of released intracellular nucleases and the subsequent isolation of intact nucleic acids. Following digestion, degraded protein is routinely removed by phenol extraction. For example, proteinase K has been used to degrade protein during the isolation of high-mol-wt eukaryotic DNA for cloning in phage or cosmid vectors, to remove protein during plasmid and lambda phage DNA isolation, and to remove protein from protein DNA complexes produced during DNA footprinting analysis. The proteinase K method for extracting RNA is a well established method. Because of its broad specificity and its ability to function in the presence of detergent, proteinase K has also been used extensively to study membrane protein topology and protein translocation across membranes. These various uses of proteinase K are further discussed in Section 3. The use of proteinase K to degrade polypeptides to produce a mixture of oligopeptides 2-6 residues in length that is suitable for sequence analysis by GC-MS has also been reported.

The chromosomal gene for proteinase K has been characterised and expression of the cDNA in *E. coli* has been achieved. The crystal and molecular structure of the enzyme have also been determined by X-ray diffraction studies to 0.15 nm resolution.

ENZYME DATA

Specificity

The enzyme is an endoprotease having a fairly broad specificity but with a preference for cleavage of peptide bonds C-terminal to aliphatic and aromatic amino acids, especially alanine. Although commonly used for its broad specificity, the enzyme has also been used to generate a single cleavage in a native protein. Lebherz et al. used proteinase K to cleave an isoenzyme of creatine kinase at a single position between two alanine residues. X-ray and model building studies on the specificity of the active site of proteinase K have been described.

Molecular Mass

Early work based on gel filtration studies suggested a mol mass of 18,500 ± 500 Da. Later studies using SDS gel electrophoresis gave values of 27,000 Da. The gene sequence was later determined and gave a true value for the mol mass of proteinase K of 28,930 Da.

Isoelectric Point

The isoelectric point of the protein occurs at pH 8.9.

pH Optimum

The pH activity curve of proteinase K, determined for the hydrolysis of urea-denatured hemoglobin, showed optimal activity in the pH range 7.5-12.0. However, the enzyme is normally used in pH range 7.5-9.0.

Assay

The assay is based on the hydrolysis ofN-acetyl-L-tyrosine ethylester (ATEE). One unit will hydrolyze I pmol of ATEE per minute at pH 9.0 and 30°C in Tris-HCI buffer, pH 8.0. Hydrolysis is monitored at 237 nm. Using this assay, the enzyme is usually supplied with a specific activity of approx 300 U/mg. Other suppliers use an assay whereby 1 Anson unit liberates 1 mmol of Folin-positive amino acids per minute at pH 7.5 and 35°C using hemoglobin as substrate. In this case, the enzyme is supplied with a specific activity of approx 20-30 U/mg of protein. Alternatively, an assay based on the hydrolysis of Suc-$(Ala)_3$NH-Np can be used. The assay is performed in 50 mMTris-HCI, 5 *mM* $CaC1_2$ (pH 8.0) using an enzyme concentration of 5-8 pg/mL and 1 mM substrate in a final volume of 1 mL. The reaction mixture is incubated for 60 min at 20°C and then stopped by the addition of 0.2 mL of glacial acetic acid. After a further 15 min, the absorbance at 410 nm of free nitroaniline is measured. Using this assay, the enzyme shows a specific activity of approx 13 U/mg. One unit of activity is the amount of enzyme that liberates 1 mmol of p-nitroaniline per minute of reaction.

Stability

Proteinase K is an unusually stable enzyme. Studies have shown that the enzymic activity of proteinase K is controlled by calcium. Hence, proteinase K is normally used in the presence of approx 2.55 mM $CaC1_2$. If calcium is removed by EDTA, followed by gel filtration, enzymic activity drops to 20% of its original value within 6 h, without autolysis. Addition of excess calcium results in an immediate rise of residual activity to 28% of the original value but full activity is not restored. The activity of the calcium bound enzyme is at a maximum at 37°C. However, the enzyme demonstrates a broad temperature profile with >80% of its maximum activity being retained between 20 and 60°C. Autolysis of the enzyme occurs during sample preparation for SDS gel electrophoresis and at low concentrations (0.01 mg/mL) in aqueous solution. It does not occur at concentrations >1 mg/mL. Proteinase K is fully active in 0.5% (w/v) SDS and is frequently used in the presence of SDS. It is also active in 1% (w/v) Triton X-100.

Stock solutions of the enzyme are normally prepared in water at 1025 mg/mL and stored at -20°C. However, protemase K may be stored in 50 mM Tris-HCl, pH 8.0, containing 1 mM $CaC1_2$ and is stable for at least 12 mo at 4°C (28).

Inhibitors

Proteinase K appears to be a serine protease, being inhibited by diisopropyl phosphofluoridate (DFP or Dip F) and phenylmethanesulfonyl fluoride (PMSF). PMSF is normally added to a final concentration of 5 *mM* from a 100 mM dry stock solution in DMSO or isopropyl alcohol. It is not, however, inhibited by tosyl lysyl chloromethyl ketone

(TLCK or Tos-Lys-CH_2CI), an inhibitor of trypsinlike serine proteases, or tosyl phenylalanyl chloromethyl ketone (TPCK or Tos-Phe-CH_2Cl), an inhibitor of chymotrypsin-like serine proteases. It is not inhibited by sulfhydryl reagents.

EXPERIMENTAL PROCEDURES

Nucleic Acid Isolation

A stock solution of the enzyme is diluted typically to 50-200 μg/mL in the solution to be digested, normally in the pH range 7.5-8.0 and at 37°C. Incubation times vary depending on the nature of the experiment, but can range from 30 min to 18 h. Although inhibitors of proteinase K are known they are generally not used as the proteinase K is usually denatured by subsequent phenol extractions. Nucleic acid purification protocols involving the use of proteinase K are described in detail in earlier volumes of this series.

Protein Studies

When studying protein translocation, membrane topology, and enzyme compartmentation, proteinase K is added to membrane preparations (or in vitro translation systems supplemented with membrane vesicles) to degrade those portions of the proteins that are accessible. Thus, cytoplasmic proteins are completely degraded, completely translocated proteins should be fully protected, whereas transmembrane proteins and those that cross the membrane more than once will be degraded to yield discrete fragments. Following proteolytic digestion, proteins are analyzed by SDS gel electrophoresis and patterns compared before and after proteolysis to identify protected polypeptides.

However, misleading results can be obtained if exposed regions of the polypeptide are inherently protease resistant. To differentiate genuine protection from protease activity and inherent protease activity, a further digestion is carried out in the presence of a nonionic detergent (e.g., 0.5-2% Triton X-100), which disrupts membranes allowing access of all proteins to proteinase K. Any polypeptide or fragment remaining after this treatment must be inherently protease resistant. The ability of proteinase K to function in the presence of detergents, and its broad specificity, makes the enzyme a very appropriate one for this type of study. Digestion of membrane is carried out at a range of concentrations (generally 20-1000) g/mL) to find optimum conditions. Proteinase K is normally prepared fresh as a 10X stock solution in 100 mm $CaC1_2$. Following digestion, it is necessary to inhibit the enzyme to prevent further degradation of proteins following protein extraction for SDS gel electrophoresis.

CARBOXYPEPTIDASE Y

Carboxypeptidases are proteolytic enzymes that remove L-amino acids, one residue at a time, from the carboxyl terminus of polypeptide chains, i.e., they are exoproteases. A number of such enzymes have been isolated from plant and animal sources, each differing in their chemical and physical properties and the rate at which they release particular amino acids. The major use of carboxypeptidases in molecular biology is in the determination of the C-terminal amino acid sequence of peptides and proteins (no suitable chemical method exists for the sequential removal of C-terminal amino acids from a polypeptide). The protein or peptide being analyzed is digested with carboxypeptidase and aliquots removed at timed intervals, and analyzed for the presence of free amino acids. The amount of each amino acid released is plotted against time, and the C-terminal sequence deduced from the relative rate of release of each amino acid. Four carboxypeptidases have been used extensively to provide peptide and protein sequence data. These are: carboxypeptidase A from bovine pancreas, carboxypeptidase B from porcine pancreas, carboxypeptidase C from orange leaves, and carboxypeptidase Y from yeast. Historically, carboxypeptidases A and B were the first to be discovered and used for sequence determination. Carboxypeptidase A releases most C-terminal amino acids, but will not cleave at arginine, proline, and hydroxyproline, or lysine residues. The specificity of carboxypeptidase B is far more restricted, cleaving only C-terminal arginine and lysine residues. Carboxypeptidases A and B, therefore, tended to be used together, but even so, exopeptidase activity was effectively blocked when a proline residue was reached. The isolation of carboxypeptidase C provided an enzyme that combined the specificity of carboxypeptidases A and B, but also cleaved at Pro residues, i.e., it cleaves all C-terminal amino acids. Carboxypeptidase Y (CPY, isolated from baker's yeast) has the same broad specificity as carboxypeptidase C, but because of its strong action on protein substrates and its ability to work in the presence of urea and detergents, it is nowadays the enzyme of choice for C-terminal sequence analysis and will be described in this chapter.

THE ENZYME

Purification

Large-scale purification of the enzyme from baker's yeast has been described by a number of workers.

Specificity

The enzyme cleaves all L-amino acids one residue at a time from the C-terminal of polypeptide chains. However, the rate of release of individual amino acids varies. Catalysis is maximum when the penultimate and/or terminal residues have aromatic or aliphatic side chains. When glycine or aspartic acid is in the terminal position, or lysine and arginine in the penultimate position, the release of the amino acid is slow. The cleavage of tripeptides is difficult, and dipeptides are completely resistant to hydrolysis. C-terminal proline is a good substrate, but a proline residue on the carboxy terminal side of glycine is not likely to be released.

The enzyme is a serine carboxypeptidase having a strongly nucleophilic serine residue at the active center, which is generated by a charge relay system involving a histidine residue. It has strong esterase activity toward the substrates of chymotrypsin and also anilidase activity. It therefore seems to be quite similar to chymotrypsin in both mechanism and active site, although carboxypeptidase Y is an exopeptidase and chymotrypsin an endopeptidase.

Physical and Chemical Properties

CPY has a mol mass of 61,000-63,000 Da as determined by SDS gel electrophoresis. It is a glycoprotein, having a single polypeptide chain of about 430 residues with 16 residues of glucosamine in the carbohydrate moiety. About 96%, i.e., 416 of approx 430 amino acid residues, have been sequenced with tentative assignment of the carbohydrate attachment sites, and the molecule shown to contain one sulfhydryl group, either four or five disulfide bridges, and five methionine residues.

pH Optimum

The optimum pH for hydrolysis of acidic amino acids is pH 5.5, while that for neutral and basic amino acids is pH 7 (5,14).

Assay

The substrate N-carbobenzoxy-Phe-Ala (N-CBZ-Phe-Ala) is prepared as a 100-mM stock solution in methanol and diluted when required in 50 mM MES buffer, pH 6.75, to give a substrate solution that is 2 mM in N-CBZ-Phe-Ala and contains 2% methanol. Ten micrograms of enzyme are added to 3 mL of substrate solution, and the decrease in absorbance at 230 nm measured with time. One unit is defined as the amount of enzyme that will hydrolyze 1.0 pmol of N-CBZ-Phe-Ala to N-CBZ-L-phenylalanine and L-alanine/min at pH

6.75 at 25°C (E^{M}_{230} = 191.5). Using this assay, the enzyme is normally provided with a specific activity of 100-130 U/mg of protein. Esterase and anilidase assays have also been well documented (10).

Stability

The lyophilised enzyme is stable for 6-12 mo if stored at 4°C. A suspension of the enzyme in saturated ammonium sulfate can be stored at -20°C indefinitely. When dissolved in water and dialysed against water to give an approx 1 % solution of the enzyme, this solution can be aliquoted and stored at -20°C for at least 2 yr. Diluted solutions (<0.1 mg/mL) lose activity fairly quickly and should therefore be prepared just before use. The enzyme is fully stable in the presence of 10% methanol at pH 5.5-8.0 for 8 h at 25°C, and in 20% methanol, 80% activity remains after incubation at pH 7 for 24 h. Enzyme activity is rapidly lost below pH 3 or above 60°C.

Denaturing Agents

Eighty percent of activity remains after incubation with *6M* urea at 25°C for 1 h. The enzyme also retains its activity for extended periods in 1 % SDS. Its stability to denaturing agents makes CPY very suitable for studying proteins that have inaccessible or poorly accessible C-termini under normal (nondenaturing) conditions.

Inhibitors

The enzyme is a serine protease, and is therefore inhibited by DFP and PMSF, but is not inhibited by soy bean and lima bean trypsin inhibitors. It is inhibited by p-hydroxymercuribenzoate, probably by reaction with a single thiol group thought to be located near or at the substrate binding site. Enzyme activity is affected by metal ions: Cu^{2+}, Ag^{+}, and Hg^{2+} result in complete loss of activity at $10^{-4}M$, 1 mm Cu^{+}, Mg^{2+}, Ca^{2+}, Ba^{2+}, Cr^{2+}, Mn^{2+}, Fe^{2+}, Fe^{3+}, CO^{2+} or Ni^{2+} results in loss of more than 50% activity. Certain organic solvents, such as DMF and ethanol, apparently show competitive inhibition. EDTA and o-phenanthroline have no effect on enzyme activity.

Additional Comments

Commercial preparations may contain free amino acids owing to autolysis, which should be removed before use. Repeated freeze-thawing of solutions of the enzyme or prolonged storage at room temperature can also lead to autolysis and the liberation of free amino acids. For ammonium sulfate suspensions, centrifuge and wash the pellet in saturated ammonium sulfate before dissolving in buffer.

Alternatively, the dissolved enzyme can be dialysed against the pyridine acetate buffer used for digestion.

EXPERIMENTAL PROCEDURE

Determination of C-Terminal Sequences in Peptides and Proteins

Dissolve 20 nmol of the protein to be studied in 200 pL of digestion buffer (0.1M pyridine acetate, 0.1 mM norleucine, pH 5.6, containing 1 % SDS). The norleucine is used as an internal standard for amino acid analysis to allow for compensation of any handling losses or sampling errors. Heat the solution to 60°C for 20 min to denature the protein. After cooling, remove a 25-μL aliquot as the zero time sample. Add 2 nmol of carboxypeptidase Y in 5-10 μL of 0.1M pyridine acetate buffer, pH 5.6, thoroughly mix, incubate at room temperature, and remove 25-μL aliquots at $T=$ 1, 2, 5, 10, 20, 30, and 60 min. Add 5 μL of glacial acctic acid to each sample to stop the reaction. Samples are then frozen, lyophilised (pyridine acetate is volatile), and then subjected to amino acid analysis. The small amount of SDS in each sample applied to the amino acid analysis does not interfere with the elution profile or affect the integrity of the machine. The sampling times indicated here should be appropriate for most proteins. However, should the C-terminal sequence be such that a number of slowly released amino acids are present, then the experiment may have to be repeated using longer incubation times or a higher enzyme-to-substrate ratio.

If analyzing peptides, SDS can be omitted from the incubation buffer, since the C-terminal should be readily accessible. In addition, the time intervals for sampling can be reduced, since the rate of appearance of free amino acids will be faster than that for proteins. Use T= 0, $^1/_2$, 1, 2, 5, 10, 15, and 20 min. A graph is plotted showing the amount of each amino acid released with time and the C-terminal sequence deduced from the relative rate of release of each amino acid.

AMINOPEPTIDASES

Aminopeptidases are proteolytic enzymes that remove L-amino acids sequentially from the amino termini of polypeptide chains. A number of aminopeptidases have been isolated, including leucine aminopeptidase from serine kidney cytosol, aminopeptidase P from *E. coli,* proline iminopeptidase from *E. coli*, and swine kidney, aminopeptidase B from rat liver, and aminopeptidase A from rat kidney.

However, three aminopeptidases in particular have found routine use in protein chemistry. The first is pyroglutamate aminopeptidase, a thiol exoprotease that cleaves N-terminal pyroglutamyl residues (pyrrolidone carboxylic acid) from peptides and proteins. N-terminal glutamine residues can readily cyclize to the pyroglutamyl derivative. This can occur during peptide and protein purification (it is uncertain whether the N-terminal pyroglutamyl residues of a number of naturally occurring peptides and proteins are genuine posttranslational modifications, or were introduced by cyclisation of N-terminal glutamine during purification) or during sequence determination when glutamine was the newly liberated N-terminal amino acid. This cyclised derivative does not have a free amino group, and therefore, the peptide or protein is not amenable to sequence determination, unless the pyroglutamyl derivative is removed by pyroglutamate aminopeptidase. The enzyme was first purified from *Pseudomonas fluoresens*, but nowadays the calf liver enzyme is used, and it is this enzyme that we describe.

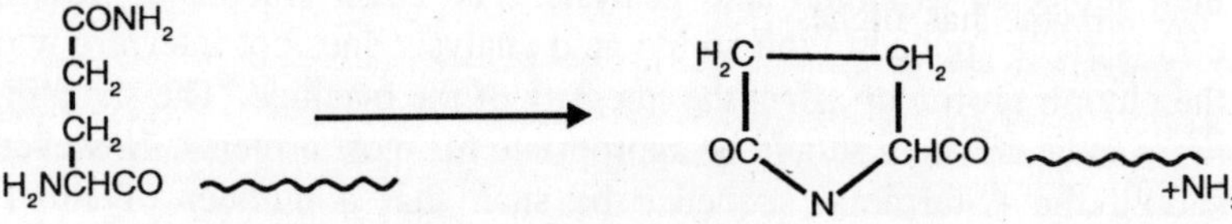

$CONH_2$ | CH_2 | CH_2 | H_2NCHCO ∿∿∿ ⟶ H_2C—CH_2, OC, CHCO ∿∿∿, N + NH_3

Figure 16.1 : The cyclization of N-terminal glutamine to pyroglutamic (pyrrolidone carboxylic) acid.

Aminopeptidase M, a zinc-containing metalloprotease, from swine kidney microsomes removes amino acids sequentially from the N-terminals of peptides and proteins. It, therefore, has some use in the determination of N-terminal sequence data, although the Edman degradation would probably be the method of choice for most workers. Aminopeptidase M is more frequently used in the preparation of peptide and protein hydrolysates for amino acid analysis. Traditionally, peptides and proteins are hydrolysed in 6NHC1, but this approach results in the total loss of tryptophan, partial loss (5-10%) of serine and threonine, and hydrolysis of asparagine and glutamine to the corresponding acids. The use of enzymes to produce a peptide/protein hydrolysate overcomes these problems. Although aminopeptidase M is capable of cleaving all possible peptide bonds, in practice, the X—Pro bond is not completely cleaved and the dipeptide X—Pro is released.

Aminopeptidase M therefore tends to be used in conjunction with our third enzyme, prolidase. Prolidase, a manganese-containing metalloprotease, has been purified from a number of sources, but the porcine kidney enzyme is generally used. It is more correctly a highly specific imidopeptidase, since it cleaves the dipeptide X-Pro or XHypro. The amino acid sequence and gene location of human prolidase have been elucidated, and active site modeling studies of the enzyme have been performed.

ENZYME DATA

Pyroglutamate Aminopeptidase (Calf Liver)

Alternative Names

These are L-pyroglutamyl peptide hydrolase, 5-oxoprolyl-peptidase, pyrrolidonyl peptidase, pyroglutamate aminopeptidase, pyroglutamyl peptidase, and pyroglutamase. *Note:* This enzyme is now classified as EC 3.4.19.3. In earlier literature, the enzyme was classified as EC 3.4.11.8.

Specificity

The enzyme cleaves N-terminal pyroglutamyl residues from peptides and proteins, but not if the following residue is proline. The enzyme has highest specificity when the pyroglutamate residue is linked to alanine.

Molecular Mass

A mol mass of 79,000-80,000 Da has been reported for the enzyme, based on gel filtration studies.

pH Optimum

The enzyme is active at pH 7-9, but is normally used at pH 8.0.

Assay

The assay is based on the hydrolysis of L-pyroglutamic acid β-naphthylamide. The enzyme buffer used is prepared as follows: 1 L of 0.1M Na_2HPO_4 is adjusted to pH 8 with 0.1M NaH_2PO_4 to give 1605 mL, and this solution is made 5% (v/v) in glycerol. A deblocking buffer is prepared by making 112 mL of the enzyme buffer, 5 mM in DTT, and 10 mM in Na_t EDTA, adjusting the solution to pH 8 and purging with N_2. Enzyme solution (0.4 mL) and 0.5 mL of deblocking buffer are combined, purged with argon, and sealed with a cap. The mixture is incubated for 3 min at 37°C, the substrate added (1.78 mmol pyroglutamyl-β-naphthylamide in 0.1 mL of methanol), and the mixture again incubated for 5 min at 37°C. One milliliter of 25% TCA solution is then added to stop the reaction. To quantify the 0-naphthylamine released, 1 mL of this solution is mixed with 1 mL of 0.1 % $NaNO_3$. After 3 min, 1 mL of a 0.5% ammonium sulfamate solution is added to destroy excess nitrite. After 5 min, 2 mL of

0.05% N-1-naphthylethylenediamine dihydrochloride solution is added, and the solution incubated at 37°C for 1 h. The solution is then cooled, and the absorbance recorded at 570 nm. The amount of β-naphthylene can be determined using a standard curve. Using this assay, the enzyme is normally supplied with a specific activity of 100-350 U/mg protein, where 1 U hydrolyzes 1 nmol of L-pyroglutamic acid 0-naphthylamide to L-pyroglutamic acid and β-naphthylamine/min. The enzyme has also been assayed using (Pro-^{3}H)-thyroliberin as substrate, and by HPLC methods that use peptides (containing pyroglutamate residues) as substrates.

Stability

The lyophilised enzyme is stable at 4°C for months. The enzyme is generally unstable. Its stability is enhanced by sucrose and EDTA in some commercial preparations. It may be reconstituted in solutions containing 5 mMDTT and 10 mMEDTA, and stored at-20°C, or it can be used for a maximum of 1 wk, if stored at 4°C. The enzyme is stable in 1 M urea and 0.1M guanidine hydrochloride. Podell and Abraham (26) have found the enzyme to be extremely unstable above room temperature, and it was found that deblocking did not occur at 37°C. It was found that an initial incubation at 4°C followed by a second incubation at room temperature was necessary to ensure maximum enzyme stabilizing conditions. Air oxidation causes severe reduction in enzyme activity, and therefore, incubations involving the enzyme should be carried out under nitrogen.

Activation/Inhibition

Podell and Abraham have studied factors affecting the stability of the enzyme and have developed a buffer (deblocking buffer) that is compatible with the use of the enzyme. The enzyme requires a thiol compound for activation and is inactivated by thiol-blocking compounds, such as iodoacetamide, and divalent metal ions, such as Hg^{2+}. Activity can be restored by short incubations with mercapto-ethanol. Dithiothreitol and EDTA are included in the deblocking buffer to overcome inactivation.

Aminopeptidase M

Alternative Names

These are amino acid arylamidase, microsomal alanyl aminopeptidase, and α-aminoacyl peptide hydrolase.

Specificity

The enzyme cleaves N-terminal residues from all peptides having a free α-amino or α-imino group. However, in peptides containing an

XPro sequence, where X is a bulky hydrophobic residue (Leu, Tyr, Trp, Met sulfone), or in the case of an N-blocked amino acid, cleavage does not occur. It is for this reason that prolidase is used in conjunction with aminopeptidase M to produce total hydrolysis of peptides.

Molecular Mass

The enzyme has a mol mass of about 280,000 Da based on gel filtration and is composed of 10 subunits (28,000 ± 3000 Da) of two different types. The enzyme molecule contains five disulfide bridges, each of which connects two subunits. The gene has been cloned, the amino acid sequence determined, and mol mass confirmed.

pH Optimum

At substrate concentrations used for sequence work, the pH optimum is between 7.0 and 7.5, but at higher substrate concentrations, approaches 9.0.

Assay

The assay is based on hydrolysis of p-nitroanilides of amino acids, especially alanine or leucine. The enzyme is added to 0.06M phosphate buffer, pH 7.0, containing 1.66 MM L-leucine p-nitroanilide or L-alanine-p-nitroanilide to give a final vol of 2.0 mL. The increase in absorbance at 405 nm is recorded at 37°C. One enzyme unit is defined as the amount of enzyme that produces 1 μmol p-nitroanilide/min at 37°C. The enzyme is normally supplied with a specific activity of 4 U/mg protein. Alternatively, an assay based on the hydrolysis of L-leucinamide at pH 8.5 and 25°C is used. Using this assay, the enzyme is usually supplied with a specific activity of 25 U/mg protein. The hydrolysis of phenylalanyl-3-thia-phenylalanine at pH 8.2 and 25°C has also been used as the basis of an assay that distinguishes between leucine aminopeptidase and aminopeptidase M.

Stability

The lyophilised enzyme is stable for several years at -20°C. A working solution can be prepared by dissolving about 0.25 mg of protein in 1 mL of deionised water to give a solution of approx 6 U of activity/mL. This solution can be aliquoted and stored frozen for several months at -20°C. The enzyme is reported to be stable at pH 7.0 at temperatures up to 65°C, and is stable between pH 3.5 and 11.0 at room temperature for at least 3 h.

Activation/Inhibition

The enzyme is not affected by sulfhydryl reagents, has no

requirements for divalent metal ions (unlike the cytosolic leucine aminopeptidase), is stable in the presence of trypsin, and is active in 6M urea. It is not inhibited by PMSF, DFP, or PCMB. It is, however, irreversibly denatured by alcohols and acetone, and 0.5M guanidinium chloride, but cannot be precipitated by trichloroacetic acid. It is inhibited by 1,10-phenanthroline (10M).

Prolidase (Porcine Kidney)

Alternative Names

These are imidodipeptidase, proline dipeptidase, aminoacyl L-proline hydrolyase, and peptidase D.

Specificity

The enzyme is highly specific, and cleaves dipeptides with a prolyl or hydroxyprolyl residue in the carboxyl terminal position. It has no activity with tripeptides. The rate of release is inversely proportional to the size of the amino terminal residue. The enzyme's activity depends on the nature of the amino acids bound to the imino acid. For optimal activity, amino acid side chains must be as small as possible and apolar to avoid steric competition with the enzyme receptor site. The enzyme has the best affinity for alanyl proline and glycyl proline. Experimental data have also suggested that prolidase may only cleave the *transform* of the peptide bond. Active site modeling studies have also been performed.

Molecular Mass

The mol mass, determined by SDS-PAGE, is 110,000 Da for the native protein and 53,000 Da for the reduced form. The enzyme is therefore thought to consist of two chains linked by disulfide bonds. Prolidase is a glycoprotein containing about 0.5% carbohydrate.

pH Optimum

The enzyme has optimal activity at pH 6-8, but it is normally used at pH 7.8-8.0.

Assay

An assay based on the hydrolysis of glycyl-L-proline at 37°C has been described. An incubation mixture (0.5 mL) containing 0.1 mL enzyme preparation in 50 MM glycyl-L-proline, 50 mM Tris-HCI, pH 7.8, and 1 mM $MnC1_2$ is prepared and incubated for 1 h at 37°C. The reaction is then stopped with 1.0 mL of 0.45M TCA, and the quantity of proline produced measured using the Chinard colourimetric method, with L-proline standards. Chinard's reagent consists of 600 mL glacial acetic acid, 400 mL orthophosphoric acid (6M), and 25 g ninhydrin, dissolved at 70°C. One milliliter of glacial acetic acid and 1 mL of Chinard's reagent are added to 0.5 mL of reaction mixture. After 10 min at 90°C, absorbance is read at 515 nm. Dilutions should be made in 0.45M TCA. Using this assay, the enzyme is usually supplied with a specific activity of 200-300 U/mg of protein. Measurement of prolidase activity using anAla-Pro substrate, isotachophoresis, or chromatographic methods has also been described.

Stability

The enzyme is most stable at pH 6-8 in the presence of 0.01M $MnCl_2$. The lyophilised enzyme is stable for many months when stored at -20°C, and is stable for several weeks at 4°C if stored in the presence of 2 mM $MnCl_2$ and 2 mM β-mercaptoethanol.

Activation/Inhibition

Manganous ions are essential for optimal catalytic activity. The enzyme is inhibited by 4-chloromercuribenzoic acid, iodoacetamide, EDTA, fluoride, and citrate. However, if Mn^{2+} is added before iodoacetamide, no inhibition is observed.

EXPERIMENTAL PROCEDURES

Removal of Pyroglutamic Acid

This method is based on the procedure described by Podell and Abraham. The sample (10 mg) is dissolved in 10 mL of deblocking buffer (0.1M sodium phosphate buffer, pH 8.0, 5 *mM* DTT, 10 mM EDTA, 5% glycerol) and placed in a screw-top vial. Approximately 25 μg of enzyme are then added, the vial flushed with nitrogen, and sealed. Protein and enzyme are allowed to incubate at 4°C for 9 h with occasional mixing. Another 25 μg of enzyme are then added, and incubation continued under nitrogen at room temperature for a further 14 h. Since the purpose of deblocking is invariably to render the protein amenable to sequence analysis, the protein can be desalted either by dialysis against 0.05M acetic acid and then lyophilised, or by passage through an HPLC gel-filtration column in 0.1 % aqueous TFA.

N-Terminal Sequence Determination by Time-Course Hydrolysis with Aminopeptidase M

Dissolve the polypeptide (1 nmol) in 49 μL of 0.2M sodium phosphate buffer, pH 7.0. Add 1 μL (0.005 U) of aminopeptidase M solution, and incubate at room temperature. Timed aliquots (5 μL) are then removed (0, 15, 30, and 45 min, and so on) and analyzed by amino acid analysis for the presence of free amino acids. The amount of each amino acid released is plotted against time, and the order of amino acids deduced from the graph.

The Use of Aminopeptidase M

Dissolve the polypeptide (1 nmol) in 24 μL of 0.2M sodium phosphate buffer, pH 7.0. Add 1 .tg of aminopeptidase M (1 μL), and incubate at 37°C. For peptides containing 2-10 residues, 8 h are sufficient for complete digestion. For larger peptides, a further addition of enzyme after 8 h is needed, followed by a further 16-h incubation. For polypeptides containing more than 35 residues, digestion with aminopeptidase M alone is insufficient. The polypeptide is first digested with the nonspecific protease pronase (I%, v/v), followed by aminopeptidase M.

Since in many cases the X-Pro- bond is not completely cleaved by these enzymes, to ensure complete cleavage of proline containing polypeptides, the aminopeptidase M digest should be treated with 1 μg of prolidase for 2 h at 37°C before analysis. The use of two or more proteolytic enzymes to produce hydrolysis will often lead to an increase in the level of background amino acids. A digestion blank should therefore also be analyzed in order to correct for the background amino acids.

ALKALINE PHOSPHATASE

Alkaline phosphatases (or alkaline phosphomonoesterases) catalyze the hydrolysis of phosphate monoesters of a variety of alcohol moieties, being most active at an alkaline pH. The enzymes have been isolated from a variety of sources, including bacteria, fungi, invertebrates, fish, and mammals (being located in many organs, including bone marrow, kidney, placenta, and intestinal mucosa), but have not been isolated from higher plants. The most commonly used alkaline phosphatases (AP) are those from calf intestinal mucosa (called CLAP, CIP, or CAP) and from the bacterium *Escherichia coli* (BAP).

The in vivo function of the enzyme is unclear, since its phosphatase activity is nonspecific. In bone tissue, it may be involved in ossification by formation of calcium phosphate, whereas in other mammalian tissues, a role in phosphate transport has been suggested.

The bacterial enzyme is expressed constitutively in most species, an exception being *E. coli,* where synthesis is repressed by orthophosphate. On phosphate starvation, BAP is secreted into the periplasmic lumen. Again a role has been suggested in a phosphate transport system.

The enzyme has two common uses in molecular biology. First, it is used as a reporter in detection systems for particular protein or nucleic acid molecules, and second, to dephosphorylate the termini of nucleic acids enabling subsequent in vitro modification.

Detection systems involve the use of an alkaline phosphatase molecule conjugated to a second protein that specifically recognizes the target molecule. This protein may be a ligand, such as streptavidin, which would bind to a biotin-labeled target, or an antibody that offers more versatility. The presence of the target-detector-enzyme complex is revealed by the action of alkaline phosphatase on a chromogenic substrate (such as 5-bromo-4-chloro-3-indolylphosphate/Nitroblue tetrazolium chloride, or BCIP/NBT), or a substrate that will produce luminescence when dephosphorylated (e.g., 3- [2'-spiroadamantane]4-methoxy-4-[3-phosphoryloxy]-phenyl-1, 2-dioxetane, orAMPPD).

Such systems enable the detection of specific proteins by ELISA or western blotting, or of DNA sequences either directly by detection of haptens that have been linked to DNA probes (e.g., biotin, digoxygenin) or indirectly by detection of the expression products of cloned genes. Specific DNA molecules can also be detected by use of oligonucleotides that have been conjugated directly to alkaline phosphatase. The numerous variations and intricacies of these AP-linked systems are beyond the scope of this chapter, which will now concentrate on the direct action of alkaline phosphatases on phosphorylated nucleic acids.

The major application of alkaline phosphatase in molecular biology is in dephosphorylating 5' termini of DNA or RNA to prevent selfligation. This is normally performed on vector DNA to reduce the number of nonrecombinant molecules produced during cloning, but may also be applied to the DNA to be cloned to prevent joining of small noncontiguous fragments that would then give spurious products.

Dephosphorylation also enables subsequent tagging with radiolabeled phosphate using the enzyme T4 polynucleotide kinase. This approach can be used for DNA or RNA sequencing, and for fragment mapping.

THE ENZYME

Bacterial Alkaline Phosphatase (BAP)

BAP is a dimer of identical or very similar subunits. The mol mass has been determined as 67,000-110,000 Da, varying with pH and ionic strength owing in some measure to tetramer formation and also transition to a random coil at low pH. The enzyme is often regarded as a compact sphere of mol mass 80,000 Da. The sedimentation coefficient of BAP, $S°_{20,w}$, is 6.0 at pH 8.0 and its isoelectric point pH 4.5.

Isozymes have been detected-normally three, but more have been reported. Active dimers can be formed from monomers derived from different sources.

The dimeric form of the enzyme contains two atoms of zinc, which are required for activity. The dimer dissociates to monomers at pH <3.0 with release of zinc ions, and chelating agents can also remove the zinc atoms, but without the formation of monomers in this case. Some workers report four atoms of zinc per dimer, only two being required for activity, and inactivation of the enzyme by metal chelators is biphasic. There is one active site per dimer at low substrate concentrations (<0.1 mM) and two at higher concentrations (> 1 mM).

Calf Intestinal Alkaline Phosphatase (CLAP)

CLAP is a glycoprotein with a mol mass of 100,000-140,000 Da, comprising two identical or similar subunits and containing four atoms of zinc per dimer. Isozymes have been found of placental alkaline phosphatase, but not CLAP. The enzyme has an isoelectric point of 5.7 independent of temperature (in the range 15-25°C) and ionic strength (0.02-0.5).

ENZYMIC REACTION

Substrate

Substrates for alkaline phosphatase are varied ranging from the phosphate esters of primary and secondary alcohols, sugar alcohols, and phenols to nucleotides and nucleic acids. This nonspecificity is reflected in the similar rates of hydrolysis of a wide range of substrates. Trans-phosphorylations also occur, often to alcohol moieties in the buffer other than water, such as Tris or ethanolamine.

Of interest to the molecular biologist is the action on the 5'-phosphate group of single- or double-stranded DNA or RNA that yields a 5'-hydroxyl group. The enzyme exhibits further activity on RNA and can hydrolyze 5'-di- and triphosphate groups, and 3'-phosphate groups. It can also cleave 2-', 3'-, and 5'-phosphates of mononucleosides. Only phosphate monoesters are susceptible and diesters are not reactive, although some alkaline phosphatase preparations will also act as pyrophosphatases. This activity can be prevented by the inclusion of Mg^{2+} in the reaction buffer, which stimulates the monoesterase activity and almost completely inhibits pyrophosphatase activity.

Temperature

The enzymic reaction occurs over a wide range of temperature, and is usually employed within a range dependent on the substrate. The rate of reaction at 37°C is twice that at 25°C.

pH

As its name implies, the enzyme is active and stable in mildly alkaline solution (pH 7.5-9.5). It is rapidly inactivated at acid pH (e.g., pH 5), because of the loss of Zn^{2+}. Lost activity cannot be restored by addition of zinc ions, but this inactivation is prevented by presence of inorganic phosphate. The pH optimum of CLAP increases with substrate concentration and decreases with increased ionic strength.

Cations

Zinc ions are essential for the activity of alkaline phosphatase. Co^{2+}, Mg^{2+}, Mn^{2+} or Hg^{2+} have been reported to substitute for Zn^{2+} in some cases. An unusual bacterial alkaline phosphatase (not E. *coli* BAP) has been described that is inhibited by Zn^{2+} ions (0.1 *mM)* and instead requires 5 mM $CaCl_2$ for activity.

In the case of CLAP, Ca^{2+}, Ni^{2+}, and Cd^{2+} have minimal effect, and Be^{2+} is inhibitory. Zinc itself has been reported to be inhibitory at high concentrations, the kinetics of which are complex, and the inhibition is not observed in the presence of glycine. The inhibition can also be overcome by the presence of magnesium. It has been suggested that magnesium can stimulate the action of additional active sites on the enzyme molecule.

Inhibitors

Inorganic orthophosphate, thiophosphate, and arsenate are strong competitive inhibitors of alkaline phosphatase with low K_i values ranging from 0.6-20 μM depending on the buffer employed. Other inhibitors of CLAP include borate, carbonate, pyrophosphate, iodosobenzoate, and iodoacetamide anions, and urea.

Metal-binding agents that complex with the zinc atoms can also inhibit, including EDTA, cyanide, α,α'-dipyridyl, and 1,10-phenanthroline.

Incubation with Zn^{2+} can reverse some of the inactivation by these agents, but Mg^{2+} and Co^{2+} are much less effective.

Many amino acids are weak inhibitors of CLAP, including L-phenylalanine and L-tryptophan, whereas D-phenylalanine has no effect. Inhibition by glycine, cysteine, and histidine is probably owing to chelation of the zinc ions. Diisopropylfluorophosphate, an inhibitor of other serine hydrolases, has only a slight effect on alkaline phosphatase (at 1-10 mM). Finally, the enzyme is also inhibited by high substrate concentrations (millimolar amounts), the precise mechanism of which is unclear.

Sulfhydryl Reagents

The presence of sulfhydryl reagents is not required for reaction. Alkaline phosphatase is reversibly dissociated by thiol reagents in the presence of urea.

Enzyme Assay and Unit Definition

One unit of alkaline phosphatase is defined as that amount of enzyme that will hydrolyze 1 pmol of 4-nitrophenyl phosphate/min. However, other conditions can vary with suppliers.

The assay is normally performed in 1M diethanolamine buffer, including 10 mM 4-nitrophenyl phosphate and 0.25 mM $MgC1_2$. However, a variety of conditions have been employed resulting in assays at pH 8.0, pH 9.6, pH 9.8, or at pH 10.5. The assay temperature has also been defined as 37°C or as 25°C. The variation in unit definitions is illustrated by the fact that 5 U measured in Diethanolamine buffer at 37°C are equivalent to I U in glycine/NaOH at 25°C, and the presence of 1M diethanolamine in a buffer can double the observed reaction rate. This plethora of unit definitions illustrates the wisdom of titrating the quantity of alkaline phosphatase required for any particular operation. A functional unit of activity is sometimes defined as being that quantity of enzyme that will dephosphorylate 1 pg (or 1 pmol) of a particular DNA species in 1 h.

EXPERIMENTAL PROCEDURES

Storage and Stability

The purity of alkaline phosphatase preparations may vary widely, and this has effects on the stability of the enzyme, as well as contaminating catalytic activities. Commercial preparations are usually assayed to be free of such activities as DNA endo- and exonucleases, RNase, Protease, and adenosine deaminase.

BAP is commonly available at 30-40 U/mg protein (with the proviso mentioned above), and CLAP at up to 900 U/mg protein. The enzyme is supplied at concentrations of 1-50 U/pL, often in a buffer containing 3.2M ammonium sulfate (65% saturated), 1 mM

$MgCl_2$, and 0.1 mM $ZnCl_2$. Desalting to remove the ammonium sulfate if present is necessary prior to reaction.

It is possible to dilute the enzyme for a short time in reaction buffer, but for longer term storage, the following specific buffer is recommended:

- 10 mM Tris-HC1 (pH 8.0-8.3)
- 1-5 mM $MgCl_2$
- 0.1-0.2 mM $ZnCl_2$
- 50% Glycerol

Triethanolamine (30 mM, pH 7.6) may be used as an alternative to Tris-110, and optional additions include 50 mM KCl and 3 mMNaCI.

The diluted enzyme is best stored at 4°C rather than at -20°C. Under these conditions, it is stable for 6 mo. The enzyme is in fact stable for several days at room temperature in neutral or mildly alkaline solution, but is inactivated by acid.

Reaction Conditions

The following reaction conditions are largely optimised for CLAP, although most are applicable to BAP also. The major difference is that reactions with CLAP are best performed at 37°C, whereas BAP is used at 60-65°C. CLAP has largely superseded BAP as the enzyme of choice, since it can be easily inactivated after reaction because of its greater heat lability.

Alkaline phosphatase is active in a wide range of buffers, and it is often possible to add CLAP directly to restriction endonuclease buffers after restriction of the substrate DNA, since the enzyme appears to work equally well as in its own specific buffer. One can add $ZnCl_2$ to 1 mM and then proceed or even include the CLAP in the digestion mixture to act concurrently with the restriction endonuclease.

Where a prepurified DNA substrate is available, the phosphatase reaction can be performed in the following buffer: 10-50 mM Tris-HCI, pH 8.0-9.0 (or 100 mM Glycine/NaOH, pH 10.5) and 0.1-1.0 mM $ZnCl_2$. In many cases, however, the presence of $ZnCl_2$ in the enzyme storage buffer renders it unnecessary to add further zinc ions to the reaction mixture. Further optional additions include 1 mM $MgCl_2$, 1 mM Spermidine, and 0.1 mM EDTA.

The quantity of substrate and enzyme used ideally needs to be titrated for each application, but is generally within the range of 1 U/l-100 pmol termini, with the larger enzyme: substrate ratios being employed when the substrate is in the form of flush or recessed termini. The substrate is usually present at a concentration of 1-50 pmol termini in a reaction vol of 20-200 μL.

Optimum reaction temperatures and times also vary with substrate:

1. DNA carrying protruding phosphorylated termini can be dephosphorylated by reaction at 37°C for 30-60 min, with the optional addition of a further aliquot of enzyme after 30 min.

2. DNA with flush ends will react under similar conditions, although reaction at higher temperatures (e.g., 50°C) may be more efficient. Two-step incubations of 15 min at 37°C followed by addition of more enzyme and further incubation at 55°C for 45 min, or 15 min each at 37°C and 56°C followed by addition of more enzyme and a repetition of the cycle give effective results.

3 DNA with recessed termini are best treated as for flush-ended DNA, or by using BAP, higher temperatures can be employed, such as incubations at 60°C for 60 min.

4. Dephosphorylation of RNA is also performed at higher temperatures, such as 55°C for 30-60 min, possibly with an initial 15 min at 37°C.

After reaction, it is necessary to remove or inactivate the alkaline phosphatase, since this will interfere with the efficiency of subsequent ligation and transformation procedures. The reaction can be terminated by adding EDTA to 10 rnM, and then extracting with phenol. The EDTA step may be improved by heating to 65°C for 60 min or 75°C for 10 min prior to phenol extraction. Other methods of terminating the reaction involve the digestion of CLAP by proteinase K (100 gg/mL in 5 mM EDTA, pH 8.0, 0.5% SDS) at 56°C for 30 min, or the chelation of the zinc ions by heating to 65°C for 45-60 min in 50 mM EGTA, pH 8.0, or 10 mM nitrilo-tri acetic acid. In all cases, a final extraction with phenol or phenol/chloroform is recommended.

The above procedures refer to CLAP, since it is very difficult to remove BAP after reaction. A newly available bacterial alkaline phosphatase isolated from an antarctic organism can be heat killed. After 30 min at 65°C, its activity is reduced to 0.01 %.

Reaction Protocol

An example of a specimen reaction protocol is given in the following:

1. Following restriction enzyme digestion, phenol/chloroform extraction, and ethanol precipitation of the required DNA, dissolve

the linearised molecule in 10 mM Tris-HCI, pH 8.3.

2. Take an aliquot containing 1 pg DNA.

3. Add 10 pL of the following l0X reaction buffer:

- 100 mM Tris-HCI, pH 8.3
- l0 mM $MgC1_2$
- 10 mm $ZnC1_2$

4. Add water to 100 pL.

5. Add 0.01 U of CLAP (diluted from stock in the specified buffer).

6. Incubate at 37°C for 30 min.

7. Add 5 pL of 10% SDS and 5 pL of 200 mM EDTA, pH 8.0. Mix well. Heat to 65°C for 60 min.

8. Extract with phenol/chloroform.

9 Add 10 pL of 3M sodium acetate, pH 7.0, and 300 pL ethanol. Precipitate the DNA at -20°C, and resuspend in the requisite buffer for subsequent manipulations (usually ligation).

Alkaline phosphatase offers the molecular biologist another tool for the manipulation of nucleic acids. Pretreatment with the enzyme allows either the specific labeling of particular molecular species or the correct synthesis of recombinant molecules. The wide specificity of AP enables most substrates to be dephosphorylated, and by good experimental design, this can be used to direct subsequent nonspecific reactions, such as DNA ligation, to yield only the desired products. The nature of the enzyme preparation and the variations in substrate require particular reactions to be optimised, and the subsequent removal of such a potent agent requires attention, but the simplicity and effectiveness of the dephosphorylation reaction render it a very widely used technique.

POLYNUCLEOTIDE KINASE

The enzyme polynucleotide kinase (PNK, or ATP:5'-dephosphopolynucleotide 5'-phosphatase) catalyzes the transfer of a y-phosphate group from a 5'-nucleoside triphosphate moiety to a free 5'-hydroxyl of a polynucleotide such as DNA or RNA, to form a 5'-phosphorylated DNA or RNA molecule and a nucleoside diphosphate.

This activity has been identified in *E. coli* infected with bacteriophages T2, T4, or T6. No enzyme can be detected in the uninfected bacterium, but a similar kinase activity has been observed in mammalian tissues, including rat liver, calf thymus, and various cell lines including Chinese hamster lung cells and HeLa cells. The bacteriophage enzymes act on both DNA or RNA, whereas the mammalian enzymes are generally active only on DNA. Exceptions to this are the calf thymus enzyme, which has a slight action on RNA, and the HeLa cell enzyme, which is solely RNA-specific.

The in vivo role of the enzyme is possibly in maintaining DNA or RNA in the 5'-phosphorylated, 3'-hydroxylated state, which is the substrate for many reactions such as ligation and packaging.

Polynucleotide kinase has many uses in molecular biology, however they can be grouped into two classes. The kinasing activity can be used purely to modify DNA, RNA, or synthetic oligonucleotides for subsequent manipulations, or it can enable the radiolabeling of these molecules for subsequent detection in probing, mapping, or sequencing experiments.

The bacteriophage polynucleotide kinases have similar properties, that from T4-infected *E. coli* having been the most studied. In this chapter, I concentrate on T4 polynucleotide kinase as an example of these enzymes.

THE ENZYME

T4 polynucleotide kinase is encoded by the structural gene *pseT*. This gene also codes for a T4 3'-phosphatase, whose activity has been identified as residing on the same enzyme molecule. The mutant phage *pse*T 1 lacks the phosphatase activity, but has unaffected kinase activity, and is therefore often used as a source for preparing the enzyme. Like many commercially available enzymes, polynucleotide kinase has also been prepared from a recombinant overproducing strain.

T4 polynucleotide kinase is a tetramer composed of identical subunits, each consisting of 45-55% a-helix and possessing an N-terminal phenylalanine residue. The mol mass of the native enzyme has been determined as 140,000 Daby gel filtration, and 147,300 Da by centrifugation. The size of the denatured and reduced mono-mers has been measured as 33,000 Da by polyacrylamide gel electrophoresis, and 33,200 Da by centrifugation. The sedimentation coefficients, $S^{\circ}_{20,w}$, are 2.95S and 6.55S for the monomer and tetramer respectively.

By comparison, the rat liver enzyme has been determined to have a mol mass of 80,000 Da by gel filtration, and a sedimentation coefficient of 4.4S.

ENZYMIC REACTION

Reaction Catalyzed

T4 polynucleotide kinase catalyzes the transfer of they-phosphate group from a 5'-nucleoside triphosphate to the 5'-OH of an acceptor molecule. This may be a nucleoside-3'-phosphate, an oligonucleotide, or a polynucleotide. The reaction is reversible, and the enzyme will catalyze polynucleotide dephosphorylation in the presence of a nucleotide diphosphate such as ADP. Excess ADP will cause the reverse reaction to be favored.

The reverse reaction can be utilised for the exchange of labeled phosphate groups between the two substrates. This exchange reaction allows 5'-labeling of polynucleotides without prior removal of the existing 5'-phosphate group.

The associated 3'-phosphatase activity causes the hydrolysis of the 3'-phosphate group of a variety of substrate molecules, including deoxynucleoside 3'-monophosphates, deoxynucleoside 3',5'-diphosphates, and 3'-phosphorylated polynucleotides, to form a 3'-hydroxyl group and release inorganic phosphate.

SUBSTRATE

Acceptor

The 5'-hydroxylated nucleoside moiety in the reaction may comprise single- or double-stranded DNA, RNA, a synthetic oligonucleotide, a nucleoside-3'-monophosphate, or a deoxynucleoside-3'-monophosphate. The enzyme can act on any molecule terminating in a naturally occurring nucleoside. By contrast, rat liver polynucleotide kinase cannot act on RNA or oligonucleotides <10 bases in length.

When the molecule is double-stranded DNA, a protruding 5' terminus is a better substrate than a blunt or recessed end. However, by increasing the concentration of the phosphate donor (usually ATP), all 5' termini can be completely phosphorylated. Increased enzyme concentrations also cause more efficient kinasing of recessed termini.

Nicks in duplex DNA will act as substrates for reaction, but the rate of kinasing is 10- to 30-fold slower than for single-stranded DNA, or protruding 5' termini, and phosphorylation is incomplete, with only 70% being achievable even after long reaction times. Raising the ATP concentration will not promote complete phosphorylation in this case.

The size of the acceptor molecule has little effect on the rate of reaction, within the range 150-50,000 nucleotides. The K_m value for large DNA fragments released by nuclease treatment is 7.6 μM. K_m for nucleoside-3'-phosphates and oligonucleotides is 22.2-143.0 pM depending on the 5' base and the length of the oligonucleotide.

Donor

The phosphate group donor for the kinase reaction may be any nucleoside triphosphate. Although ATP is used routinely for experimental or assay purposes, CTP, UTP, GTP, dATP, and TTP perform equally well. The ATP concentration should be at least 1 pM for the reaction to proceed, and excess ATP is required for optimal kinasing.

The following K_m values have been determined for T2 polynucleotide kinase: 14 μM for ATP, 15 μM for UTP, 33 μM for GTR and 25 μM for CTP. With the T4 enzyme, the K_m value for ATP is 13-140 PM depending on the DNA acceptor.

Under the conditions for phosphate exchange, the K_{mATP} (forward reaction) is 4 μM, and the K_{mADP} (reverse reaction) is 200 μM. The optimal concentrations for exchange are 10 μMATP and 300 μMADP, but even so reaction is usually incomplete.

For the dephosphorylation of single-stranded oligonucleotides, *KmADP* is 0.22 *μM,* but again only partial dephosphorylation can be achieved.

Temperature

As with most *E. coli* bacteriophage-derived enzymes, the optimal reaction temperature is 37°C.

T4 polynucleotide kinase will also perform at lower temperatures, which is a useful attribute. The rate of the kinase reaction at C°C is reduced to 7% of that at 37°C. However, the exchange reaction rate is reduced much further, to 1.2% of that at 37°C, so under these conditions, kinasing is greatly favored.

pH

The optimum pH range for T4 polynucleotide kinase is 7.4-8.0, with maximum activity in Tris buffers being observed at pH 7.6. The reverse reaction has a pH optimum of 6.2, in imidazole buffer.

The 3'-phosphatase activity is greatest at pH 5.9.

Cations

Polynucleotide kinase has a requirement for magnesium ions for both the forward and exchange reactions, with no activity being detectable in their absence. At the optimal pH of 7.6, the optimum magnesium concentration is 10 mM.

Manganese can partially replace magnesium in some cases. Note that 3.3 mM Mn^{2+} will permit reaction at 50% of the maximum rate obtainable with 10 mM Mg^{2+}.

The 3'-phosphatase activity similarly requires Mg^{2+}, with the reaction rate falling to 2% in its absence. In this case Co^{2+} can replace Mg^{2+} to some extent.

Activators

The activity of polynucleotide kinase is stimulated by sodium chloride, potassium chloride, and polyamines. Spermine promotes tetramer formation, and potassium chloride maintains the enzyme in this oligomeric form. A total of 1.7 mM of spermine can increase reaction rate by 30-fold. Spermidine, in addition to enhancing the reaction rate, also has the ability to inhibit nucleases present in some kinase preparations.

Polyethylene glycol (PEG 8000) also improves the efficiency of the reaction. Stimulation depends on the PEG concentration, which should ideally be titrated within the range 4-10%. The stimulating action of PEG is owing to it causing the DNA to undergo a "psi" transition and collapse into a highly condensed state (28). This state is only achievable for DNA molecules >300 by in length, so PEG has little effect on shorter molecules.

Inhibitors

T4 polynucleotide kinase is inhibited by inorganic phosphate and pyrophosphate. If 70 mM sodium or potassium phosphate buffers (pH 7.6) are employed, enzyme activity is reduced to 5% of that observed in Tris buffer. The reaction rate in Tris buffer can be halved by the addition of phosphate to 7-20 mM or pyrophosphate to 5 mM. In addition, 50 mM of potassium phosphate will cause 60% inhibition of the exchange reaction.

Inorganic phosphate is, however, relatively more inhibitory to *E. coli* alkaline phosphatase than to T4 polynucleotide kinase. It can therefore be used to enhance phosphorylation when phosphatase is present, e.g., when end-labeling molecules after a prior phosphatasing step.

Another strong inhibitor of polynucleotide kinase is the ammonium ion. For example, 75% inhibition can be caused by the presence of 7 mM ammonium sulfate.

Potassium chloride has an activating effect, but high concentrations can be inhibitory. This inhibition is not observed in substrates with protruding 5'-OH termini. Sodium chloride, although stimulating kinase activity toward single-stranded substrates, can also be inhibitory with some duplexes.

Sulfhydryl Reagents

Sulfhydryl reagents are essential for the action of polynucleotide kinase, whose activity falls to 2% in their absence. Maximum activity can be obtained in the presence of 5 mM dithiothreitol (DTT). However, 80% of this maximum activity can be achieved using 10 MM β-mercaptoethanol in place of DTT, and 70% with 10 mM glutathione.

ENZYME ASSAY AND UNIT DEFINITION

Assaying the activity of T4 polynucleotide kinase entails measurement of the transfer of a radiolabeled phosphate group from γ-^{32}P ATP to an acid insoluble product. The phosphate acceptor is normally a duplex DNA molecule, enzymatically treated to create 5'-hydroxyl groups. This may be achieved by partial digestion using micrococcal nuclease, which specifically produces 5'-hydroxyl and 3'-phosphate termini, or using pancreatic DNAse followed by alkaline phosphatase, which removes the 5'-phosphate groups originally created.

The standard assay conditions are as follows: 70 mM Tris-HCl, pH 7.6; 10 mM $MgC1_2$; 5 mM DTT; 66 μMy-^{32}P-ATP; and 0.26 mM 5'-OH salmon sperm DNA. Incubate at 37°C.

Commercial suppliers of the enzyme may use slightly different assays, including variations in the nature of the DNA acceptor, the concentrations of 5' termini and ATP, and the inclusion of other buffer components such as spermidine. One unit of activity is defined as the amount of enzyme that catalyzes the incorporation of 1 nmol of ^{32}P into an acid insoluble form in 30 min at 37°C.

The 3'-phosphatase activity can be determined by incubation with AMP. The enzyme is incubated with 16 mM 3'-AMP for 60 min at 37°C, and the released inorganic phosphate measured. Preparations from mutant or recombinant sources that are nominally phosphatase-free hydrolyze $<$0.1 % of the AMP.

EXPERIMENTAL PROCEDURES

Uses of Polynucleotide Kinase

The major use of T4 polynucleotide kinase in the molecular biology laboratory is for the specific phosphorylation of the 5' termini of DNA and RNA, either by direct kinasing or by the exchange reaction. This may be for the purposes of labeling the molecule, or merely to enable the molecule to be further manipulated.

End-labeling allows the quantification of the termini, the enzyme often being used in conjunction with alkaline phosphatase to assess the number of 5'-phosphate groups present by measuring the available 5'-hydroxyl groups before and after phosphatasing. It can also be used to characterize cleavage points in nucleic acids.

End-labeled oligonucleotides can be used as primers for sequencing. Both *oligonucleotides* and *polynucleotides* can be used as hybridisation probes for clone or genomic characterisation, restriction mapping using partial digestion techniques, DNA or RNA *fingerprinting,* DNA footprinting, nuclease S 1 analysis, physical mapping, and sequence analysis.

Phosphorylation reactions where the phosphate moiety is unlabeled (or labeled purely for the purposes of monitoring the reaction) are used in the synthesis of substrates for DNA or RNA ligation. These may be vector molecules, genomic fragments, or synthetic oligonucleotides such as linkers. Such manipulations allow the assembly of long nucleic acid molecules from short synthetic precursors.

The mutant 3'-phosphatase-free enzyme is especially useful for RNA analysis since the continued presence of the 3'-phosphate group prevents cyclization of concatenation.

Finally the wild type enzyme may be used as a specific 3'-phosphatase under the right conditions.

Storage and Stability

Enzyme preparations are commercially available at concentrations of 1000-12,000 U/mL, and with specific activities in the range 30,00 040,000 U/mg.

The enzyme is normally stable for up to 18 mo at-20°C in a suitable storage buffer. The composition of this buffer can vary, but a general purpose buffer used for the storage of a wide variety of enzymes can be used, with the optional addition of 0.1-1.0 pM ATP. An example of a typical storage buffer would be: 50 mM Tris-HCI, pH 7.5; 25 mM KCI; 1 mM DTT; 0.1 mM EDTA; 0.1 pM ATP; and 50% glycerol.

Reaction Conditions

T4 polynucleotide kinase will act in restriction enzyme buffers allowing simultaneous restriction digestions, but specific conditions have also been defined as described elsewhere in this chapter. Reaction conditions can be varied to give the optimal reaction with a particular substrate. For example, when kinasing blunt or 5'-recessed termini, higher concentrations of ATP and enzyme are necessary, and additional buffer components such as PEG may be beneficial.

A number of points are worthy of note. First, because of the extreme inhibition of T4 polynucleotide kinase by ammonium salts, the preparation of the polynucleotide substrate should not involve precipitation with ammonium acetate prior to the kinasing step. Second, all reaction buffers used should be Tris- or imidazole-based, and not contain inorganic phosphate; otherwise inhibition will be observed again. Finally, the target molecule should be rigorously purified from small mol-wt fragments, since these would contribute a large number of 5' termini, which could compete in the reaction.

After the kinasing reaction, subsequent purification depends on the use to which the phosphorylated polynucleotide will be put. If it is to be used as a hybdridization probe, further purification is not necessary unless problems with background noise have been encountered. Purification by repeated ethanol precipitation may be sufficient. When working with oligonucleotides of < 18 bases in length, precipitation with cetylpyridinium bromide must be employed. If the molecule is to be used in further manipulations, it may need more rigorous purification by, e.g., column chromatography, spin dialysis, or polyacrylamide gel electrophoresis.

Reaction Protocols

Materials Required

1. Suitably prepared 5'-OH DNA, RNA, or oligonucleotides. 2. Reaction buffers (use one or the other):

a. For kinasing, 1 OX Tris kinase buffer: 500 mM Tris-HCI, pH 7.6, 100 MM $MgC1_2$, 50 mM DTT, 1 mM Spermidine, and 1 mM EDTA.

b. For kinasing or phosphate exchange, l0X imidazole buffer: 500 mM imidazole-Cl, pH 6.4; 180 mM $MgC1_2$; 50 mM DTT; 1 mM Spermidine; and 1 mM EDTA.

3. γ-^{32}P-ATP (3000 or 5000 Ci/mmol).
4. T4 polynucleotide kinase (10 U/μL).
5. 500 mM EDTA, pH 8 0.
6. Sterile distilled water.
7. Phenol/chloroform and chloroform for extraction.
8. 7.5M Ammonium acetate.
9. Ethanol.
10. TE buffer: 10 mM Tris-HC1, pH 8.0, 1 mM EDTA
11. 24% PEG 8000.
12. 1 mM ADP solution (for exchange reaction only).
13. 50 nM ATP solution (for exchange reaction only).
14. Sephadex G-50 for column or spun column purification after reaction.

15. For measurement of incorporation:

a. 10% trichloroacetic acid, 5% trichloroacetic acid, and 70% ethanol;

b. DE81 filters and 500 mM sodium phosphate, pH 7.0; or

c Polyethyleneimine cellulose TLC strips and 500 mM ammonium bicarbonate.

Kinasing Single-Stranded DNA Fragments or Duplexes with Protruding 5' Termini

1. Prepare and purify dephosphorylated DNA fragments.
2. To an aliquot containing 1-50 pmol ends add:
   - 5 μL of lOX Tris kinase buffer
   - 15 μL of γ-^{32}P ATP (50 pmol, 3000 Ci/mmol, 10 Ci/μL, 1 μM final concentration)
   - 1 μL of T4 polynucleotide kinase (10 U)
   - Water to 50 μL.
3. Incubate at 37°C for 30 min.
4. Terminate the reaction by adding 2 μL of 500 mM EDTA.
5. Purify by phenol/chloroform extraction.

6 Remove unincorporated ATP by column or spun-column chromatography using Sephadex G-50.

7. Add 1/2 vol 7.5M ammonium acetate and 2 vol ethanol. Precipitate the DNA at -70°C for 30 min. Centrifuge, drain pellet, and redissolve in 50 μL TE.

8. Determine incorporation of label by TCA precipitation.

Kinasing DNA Duplexes

1. To an aliquot of DNA containing 1-50 pmol ends in 9 μL or less, add:
   - 4 μL of lOX imidazole buffer
   - Water to 13 μL
   - 10μL of 24% PEG 8000
   - 15 μL of y-^{32}P ATP (50 pmol, 3000 Ci/mmol, 10 [tCi/ μL)
   - 2 μL (20 U) of T4 polynucleotide kinase
2. Incubate at 37°C for 30 min.

3 Terminate the reaction by adding 2 μL of 500 mM EDTA, pH 8.0.

4. Extract, purify, and measure incorporation.

Exchange Reaction

1. To 1-50 pmol DNA in a small volume add:
   - 5 μL of lox imidazole buffer
   - 5μL of 1 mM ADP
   - 1μL of 50 nM ATP
   - 15 μL of γ-^{32}P ATP (50 pmol, 3000 Ci/mmol, 10 iCi/μL)
   - Water to 38 μL
   - 10 μL of 24% PEG 8000
   - 2 μL (20 U) of T4 polynucleotide kinase

2. Incubate at 37°C for 30 min.

3. Terminate the reaction by adding 2 μL of 500 mM EDTA, pH 8 0.

4. Extract, purify, and measure incorporation.

Labeling Oligonucleotides

1. To 10 pmol of oligonucleotide in 1 μL add the following:
   - 2 pL of lOX Tris kinase buffer,
   - 5 μL of γ-^{32}P ATP (10 pmol, 5000 Ci/mmol, 10 pCi/μL), and
   - 11.5 pL of water.

2. Mix and take 0.5-μL zero-time aliquot and add this to 10 pL TE.

3. To the remainder of the reaction mix add I iL (10 U) T4 polynucleotide kinase.

4. Incubate at 37°C for 45 min.

5. Take another 0 5-μL aliquot and add to 10 μL of TE as before.

6. Terminate the reaction in the remainder of the mix by heating to 68°C for 10 min.

7. Measure incorporation efficiency as follows. Spot 0.5 μL of each diluted aliquot onto 15-cm long polyethyleneimine cellulose strips Perform thin layer chromatography using 500 mM ammonium bicarbonate as the developing solution. Allow the solvent front to run 10-13 cm. Saran Wrap and autoradiograph, or slice up the strip and measure the radioactivity along the strip by scintillation counting. Compare timezero and 45-min samples. In this TLC system, oligonucleotides remain at the origin, inorganic phosphate migrates near the solvent front, and ATP occupies an intermediate position.

An alternative method is to measure incorporation by adsorption to DE81 filters. Oligonucleotides bind tightly, whereas ATP can be washed off with 0.5M sodium phosphate, pH 7.0.

8. If the oligonucleotide is not labeled highly enough, add another 10 U of enzyme and incubate for a further 30 min.

9. Purify as required.

Note: Equal concentrations of ATP and 5' ends gives 50% labeling. To obtain high specific activity, increase the ATP: oligonucleotide ratio to 10:1. Only 10% of the label will be transferred, but virtually every oligonucleotide molecule will be labeled and to a specific activity approaching that of the ATP.

INDEX

A

A cholinergic agonist 174
absorption 1
acinar cell 45, 70
active secretion 95
addition. 166
adsorbed 156
affinities 48
aldosterone 128
algebraic sum 48
alleles 118
allelozymes 119
allostery 22
amount 30
amylase 121, 181, 226
anatomical 42
anatomically 4
antibodies 56
antisense oligonucleotides 56
Apparent disappearance 70
archaebacteria 117
aromatic 131
artifact 154
assume 72
assumption 87
assumptions 6
atomistic 42, 44
augmented secretion 97, 101
autoradiography 163
auxiliary hypotheses 8
average 147
Axonemal 57
axonemal dynein 59
Axonemal dyneins 59

B

backround 164
better activation 199
binding 12
biochemical 139, 140, 162
biochemical diversity 59
biosynthesis 128
black box 170
Brownian motion 89
budding 87
bulk 65

C

canceled" 173
carbonic anhydrase 121
carboxyl termini 62
carboxylase 113
catalytic 144
catalytically 113, 139
catalyze 113
catch 73
Cell biology 40
cell division 54
cell migration 54

cell physiology 54
centrifugal force 147
channels 65
chemical 42
chemical quench flow 81
chemically 4
cholesterol 126, 128
Cholinergic 203
chromatography 122
chromosomes 54
chymodenins 223
chymotrypsinogen
158, 201, 208, 213
chymotrypsinogen, 159
cilia 55
cisternal packaging 45
clearer 164
cocarboxylase 113
coenzyme 113
coenzymes 20, 113
cofactors 20
coincident 179
colorimetric 122
competitive 178
component 50
concentration 30
condensing vacuole 92
condensing vacuoles
45, 90, 91
conformational enzymes 125
conformers 125
Constructed hypotheses 6
constructionist 5
cortisol 128
covalently 140
cytochrome system 116
cytologic 95
cytoplasm 145
Cytoplasmic dyneins 57, 63
cytoplasmic fraction 1 47
cytoplasmic microtubules 57
cytoplasmic pool
145, 151, 160
cytoskeletal 61
cytoskeletal elements 54
cytoskeletal structur 54
cytoskeletal structure 62
cytoskeleton 57, 62

D

damages 53
Davson-Danielli model 46
deblenderizer 148
deductive hypothesis 73
dehydrogenase 118
depend upon 102
depleted 111
depletion 95, 106
depression 172
desensitisation 130
detoxification 130
diffusion coefficient 50
digestive enzyme 3, 156
direct 67, 170
disappear 71
dispersed particles 146
duct lumen 84, 101
dynamic 42
dynamic process 72
dynamical 74
dyneins 55

E

E. coli
29, 237, 243, 257, 258,
243
easier 48
electrofocusing 122
electron microscope
41, 43, 89, 147
electrophilic 114
electrophoresis 122

electrophoretic mobility 121
embryogenesis 54
embryonic stem cells 55
emissions 163
empty 93
endocytic models 40
endogenous 126, 128, 131
endoplasmic 126
endoplasmic reticulum 45
Engineered 143
enzymatic 122
enzyme 31
enzyme, 156
Enzyme Nomenclature 14
enzyme units 31
enzymoelectrophoresis 122
enzymology 54
equilibrium 15
Escherichia coli 114, 134, 250
estrogen 128
eukaryotic 130
eukaryotic cell 54
evidence 232
evidence of form 69
exocytoses 75
exocytosis
2, 39, 45, 46, 65, 69,
70, 71, 73, 193
exocytosis hypothesis 75, 98
exocytosis model 99, 192
exocytosis sites 173
exogenous 126
extracellular space 171

F

filopodia 55
fingerprinting 120, 263
first-order 20
firstorder rate constant 79
flagellar dynein 57
flavoprotein 127
formyl 117
found 166
freeze-fracture 75
function 1
functionally discontinuous 46
furiosus 27

G

genetic code 4
geometric 72
Golgi 45
Golgi complex 44, 89, 94
Golgi membrane 94
Golgi membranes 45
granule loss 106
granule number 106
granule size 106
granule volume density 100

H

helps 76
hemoglobin 126
hemoprotein 126
heterogeneity 54
heterologous 132
heteropolymeric 119
hexokinase 121
hexosaminidase 119
highly 12
histochemical 118
holoenzymes 113
homogeneous protein 146
homogenization 147
Homologies 60
homologies 60
homologous compounds 51
homopolymeric 119
hormones 1
hydrocarbons 131
hydrolyse nucleotides 55
hydrolysing nucleotides 55
hyperthermostable 25

Hypotheses 8
hypothesis 76
hypothesize 150

I

Immunological 122
immunological 130
in situ 69, 149, 210, 213
in vitro 31
in vivo 31, 56
inappropriate 150
incorrect 200
independently 6
inhibited 204
initial rate 31
insensitivity 207
intact 163
International Unit 30
intracellular 171
intracellular granules 101
intracellular structures 146
intramolecular 140
investigating 139, 143
irreversible 46
isoelectric point 159
isoenzyme 118
izos 118

K

katal 30
Kinesin 57
Kinesins 57
kinesins 55
kinetic 6, 42
kinetics 14

L

lactalbumin 223
Lactate dehydrogenase 119
leak 52
ligand 22
light microscope 40, 44
light microscopy 70
lipid bilayer 49
logical empiricism 8
loss 71
lymphocytes 55

M

macro 125
macro-CK 125
macroamylase 125
macromolecules 50
macroscopic 39
major assumption 46
malignant 119
mammalian 116
mature 92
mechanical force 54
mechanism 2
mechanochemical 56
mechanochemical enzyme 58
mechanochemical enzymes 54
membrane organelles 62
mercaptoheptanoylthreonine 117
metabolise 126
metabolism 128, 131
methanol dehydrogenase 116
methionine 114
methyl 117
methylene 117
microfilament 61
microfilaments 61, 62
micromole of enzyme 31
microorganisms 126
microscopic 39, 40
microscopy 168
microsomal 128
Microtubule 61
Microtubules 61
microtubules 57, 62, 63
migrate 89

mitochondria 54, 145
mitochondrial 116
mitochondrial membrane 127
mitochondrion 146
mobility 51
mobilized 178
molecular activity 31
molecule 139
molecules 140, 141
monotonous 140
morphogenesis 54
morphology 59
mutant phenotypes 56
myosins 55

N

natural 152, 154, 157
negative evidence 156
neurons 58
neurotransmitters 1, 69, 171
never 203
nevertheless nonspecific 159
new hypotheses 41
Newtonian models 5
nonhelical 140
nonhydrolysable 58, 61
nonisoenzymic 125
nonlinear relationship 109
nonneuronal cells 58
Nonparallel 198
nonparallel transport 205, 232, 233
nucleotide 61
nucleotide hydrolysis 55
nucleotide phosphatase 56
nucleotides 127, 140
nucleus 145

O

obligatory 139
obligatory parallel 205
oligomeric isoenzymes 119
oligonucleotides 130, 263
omega figure 71, 72, 73, 77
one micromole per minute 30
opaquing agent 163
organelles 63
organic cofactors 113
orthophosphate 56
osmotic 102
osmotic effect 104

P

P. furiosus 27, 29
P. woesei 29
packagingexocytosis 153
pairs 234
pancreas 101
paradox 48
parallel secretion 193
Parallel transport 192
parallel transport 205
particles 74
pathogen 143
Patterns of particles 74
peptide hormone 3
permeability 48, 50, 51, 177, 195
permeable 49
permissive 205
permit 47
phagocytosis 67
pharmacological agents 56
philosophical 4
phosphofructokinase 121
physiological 2
physiology 43
pictures 41
pinocytotic 40
placental alkaline phosphatase 119
polynucleotides 263
polypeptide 55, 119

pore area 52
potential differences 52
prebiotic 140
preceded 5
prevent 47
primary spermatocytes 119
prior 175
process secretion 3
progesterone 128
prosthetic 113, 126
protein capacitor 110
protein secretion 3, 88
protein synthesis 44
Protein transport 3
protein transport 3, 4
proteins 171
Protists 55
Ps furiosus 29
Pseudomonas fluoresens 244
pulse 171
pulse label 172
purpose 1, 2
putative motor proteins 56
putresine 114
Pyrococcus furiosus 26
Pyrococcus woesei 27
pyrroloquinoline quinone 116
pyruvate kinase 121

Q

Quantitative measurements 95

R

radioactive 163
rapidly 174
Real 66
real 41, 73
real world 43
reconstruct 42
redox 20
reduced 4
reducing 43
reductionism 4
reductionist 42, 44
reductionistic 69
reflection coefficient 52
relative 196, 218
release 109
replication 141
reproduce 198
require 95
RER 44
researchers 143
resolution 163
response 2
responsiveness 193
reverse pinocytosis 77
revolutionised 140
riboflavin 113
ribonucleases 141
ribosome 94, 171
ribosomes 44, 45, 155
Ribozymes 139
ribozymes 23, 143
rigorous 5, 56
rough endoplasmic reticulum 84

S

saturation 20
scientific paradigm 8
second-order rate constant 79
secreted 2
Secretion 1
secretion 1, 2, 3, 39, 198
secretion). 201
Secretory granules 69
secretory granules 69
sections 164
see the process 76
seek support 67
seeking 67
segregated 45

selfreplicating 140
semipermeability 47
sequestered 45, 84
shuttle 94
signature 55
simple diffusion 48
single exponential 80
slowest 17
smooth-surfaced vesicles 85
soluble 153
special 48
Specific activity 31
specific activity 152
spermidine 114
spliceosome 144
spliceosomes 141
steady state 21
Steady state kinetic 82
steady state. 79
Stokes-Einstein equation 50
Stokes-Einstein relationship 52
strength 68
subunits 22
sulfur-dependent 25
support 67
synaptic vesicles 54

T

T. litoralis 29
tail 62
test 67
testing a hypothesis 67
testosterone 128, 131
tetrahydromethanopterin 117
Thermococcus litoralis 27
thermodynamic 6, 42
Thermotoga 27
Thermotoga maritima 29
thiamine pyrophosphate 113
three enzymes. 222
toto. 71
transform 248
transient state 79
transmission 76
Transport 46
transport 39, 42
Tritirachium album 236
truth 67
trypsin 160
trypsin inhibitors 160
trypsinogen
158, 159, 201, 208, 226
trypsinogen secretion 210
tryptophan tryptophylquinone
117
turning-over granules 174
Turnover number 31

U

ubiquitous flavoprotein 127
unnat-ural 151
unnatural 150, 151, 152,
154, 156, 157, 159

V

vacuoles 46
vectorial 46, 84
vesicle 45
vesicle formation 88
vesicles 46
vesicular 85
vesicular transport 65, 90
fluorescence microscopy 56
volume 104, 106

Z

zymogen granule
94, 100, 105, 165
Zymogen granules 44, 70, 84,
90, 94, 157, 165, 179
zymogram 122
zymos 118